Science Fairs *Plus*

Science Fairs Plus

Reinventing an Old Favorite, K–8

An NSTA Press Journals Collection

Arlington, Virginia

Claire Reinburg, Director
Judy Cusick, Associate Editor
Carol Duval, Associate Editor
Betty Smith, Associate Editor

Cover Design by Naylor Design, Inc.
Cover Illustration by Robert Soule
Inside Illustrations by Linda Olliver

ART AND DESIGN Linda Olliver, Director
NSTA WEB Tim Weber, Webmaster
PERIODICALS PUBLISHING Shelley Carey, Director
PRINTING AND PRODUCTION Catherine Lorrain-Hale, Director
Nguyet Tran, Assistant Production Manager
Jack Parker, Desktop Publishing Specialist
PUBLICATIONS OPERATIONS Erin Miller, Manager
sciLINKS Tyson Brown, Manager

NATIONAL SCIENCE TEACHERS ASSOCIATION
Gerald F. Wheeler, Executive Director
David Beacom, Publisher

Printed in Canada by Webcom.
Printed on recycled paper.

Science Fairs Plus: Reinventing an Old Favorite, K–8
NSTA Stock Number: PB173X
02 03 04 5 4 3 2 1

Library of Congress Cataloging-in-Publication Data
Science fairs plus: reinventing an old favorite, K–8.
p. cm.
"An NSTA Press journals collection."
Includes bibliographical references.
ISBN 0-87355-219-9
1. Science projects. 2. Science—Study and teaching (Elementary) 3. Science—Study and teaching (Middle school).
II. National Science Teachers Association.
Q182.3 .S3437 2003
507'.8—dc21 2002015687

Featuring sciLINKS®—a new way of connecting text and the Internet. Up-to-the-minute online content, classroom ideas, and other materials are just a click away. Go to page ix to learn more about this new educational resource.

Contents

Project Selection

Potpourri

Section II: Expos and Festivals

About This Book

The National Science Teachers Association (NSTA) has assembled this collection of selected reprints from three of its journals—*Science and Children*, the journal for elementary teachers; *Science Scope*, the journal for middle and junior high school teachers; and *The Science Teacher*, the journal for high school teachers—to help K–8 teachers organize and conduct successful science events with their students. Whether you decide to conduct a traditional science fair or explore a science expo or festival, this book provides you with practical information to ensure success.

NSTA Position Statement on Science Competitions

The National Science Teachers Association recognizes that many kinds of learning experiences, including science competitions, can contribute significantly to the education of students of science. With respect to science competitions such as science fairs, science leagues, symposia, Olympiads, and talent searches, the Association takes the position that participation should be guided by the following principles:

I. Student and staff participation in science competition should be voluntary.
II. Emphasis should be placed on the learning experience rather than on the competition.
III. Science competitions should supplement and enhance other educational experiences.
IV. The emphasis should be on scientific process, content, and/or application.
V. Projects and presentations must be the work of the student with proper credit to others for their contributions.

—The NSTA Board of Directors adopted this position statement in July 1986. In fall 2002, the Board was in the process of revising the statement; go to *www.nsta.org/position* for the latest version.

How can you and your students avoid searching hundreds of science websites to locate the best sources of information on a given topic? SciLinks, created and maintained by the National Science Teachers Association (NSTA), has the answer.

In a SciLinked text, such as this one, you'll find a logo and keyword near a concept your class is studying, a URL (*www.scilinks.org*), and a keyword code. Simply go to the SciLinks website, type in the code, and receive an annotated listing of as many as 15 web pages—all of which have gone through an extensive review process conducted by a team of science educators.

Need more information? Take a tour—*http://www.scilinks.org/tour/*

Introduction

Volcanoes Are OK
and Other Divine Secrets of Successful K–8 Science Fairs, Expos, and Festivals

Donna Gail Shaw

After twenty years of teaching science at the elementary, middle, and college level, and almost that many years serving as a judge at local and state science fairs, I and other authors published in this book have discovered some secrets that teachers will find helpful when facing the task of organizing a science event or getting students ready for the event. Read on. These secrets are divine.

Secret 1. Organizing a school science fair is not complicated.

Just the thought of organizing a science fair for an elementary or middle school can seem overwhelming. Deborah Fort and Ruth Bombaugh have uncovered numerous ways to ensure the smooth operation of a school science fair. Fort (page 4), in her description of the start-to-finish operation of a science fair at an elementary school, offers advice on everything from deciding what a science project actually is, to picking the judges, to conducting a science fair wrap-up. Bombaugh (page 12), self-described as organizationally challenged, shares a master schedule she developed for organizing a middle school science fair that has passed the test of time.

Secret 2. Picking a topic is the easy part.

Gail Foster (page 20) states that selecting a topic and identifying a problem can be the most difficult parts of the science fair project. However, after making this statement, she lets us in on a secret: she provides expert guidance on how to make the process of selecting the topic easy for the students. Foster explains how to (1) introduce students to the idea of asking questions, (2) create excitement and arouse curiosity, (3) help students who need additional assistance in picking a topic, and (4) narrow broad topics to specific problem statements. Susan Shaffer (page 16), as well as Fort (page 4), also offer suggestions for helping students select topics. In addition, Shaffer lists Internet sites for finding help with science fair ideas and preparation. (See also Appendix A, Resource List, of this book for other science fair sites on the Internet.)

Secret 3. Judging science fair projects can be effective and fair.

While there is some debate over whether science fair projects should be judged at all (Fort, page 4; Evelyn Streng, page 38), Lawrence Bellipanni, Donald Cotten, and Jan Marion Kirkwood (page 48) share five basic criteria for judging projects. They stress the importance of the judges using the same criteria in the same way to ensure fairness. Bellipanni and James Edward Lilly (page 30) share similar criteria as well as suggestions for the selection and preparation of judges. Norman F. Smith (page 58) highlights some of the pitfalls of judging and proposes a separate judging category for experimental projects. The divine secret is found in understanding that the focus should be on learning rather than on competition; realizing the potential pitfalls; establishing clear, objective criteria; selecting appropriate personnel to serve as judges; and training the judges in the proper use of the scoring rubric using actual projects.

Secret 4. Teaching the scientific method to children of all ages is possible.

Donna Gail Shaw, Cheryl Cooke, and Teralyn Ribelin (page 40) share how to teach the scientific method to a diverse group of learners in a K, 1, 2 multi-age, full-inclusion classroom by conducting a whole class project. In accordance with what is known about child development and the brain, it should be noted that primary level students (K–3) may have difficulty manipulating more than one variable at a time (Lowery 1998). Understanding that there is a biological basis for this difficulty will lessen the teacher's frustration while guiding this age group to use inquiry-based science. Stephen Blume (page 24) simplifies the process for elementary students by suggesting the use of six easy questions. Finally, Charlene Czerniak (page 26) shares a quick way to teach middle school students the scientific method by using either quantitative or qualitative measures. The divine secret is in the way the information is presented to children of varying ages and one's belief in their ability to learn the process.

Secret 5. Conducting nonexperimental research can meet the goals for school science.

The goals for school science (National Research Council 1996) are to educate students who are able to

- experience the richness and excitement of knowing about and understanding the natural world;
- use appropriate scientific processes and principles in making personal decisions;
- engage intelligently in public discourse and debate about matters of scientific and technological concern; and
- increase their economic productivity through the use of the knowledge, understanding, and skills of the scientifically literate person in their careers. (13)

John Stiles (page 64) and Smith (page 58) deplore the current plight of elementary and middle school science fairs. Based on their experiences as science fair judges, they believe that the only worthwhile projects are experimental in nature and that models, demonstrations of principles, and report and poster projects do not promote the goals of science teaching. However, Streng (page 38) expands on the type of projects she considers appropriate by including any problem-centered projects that focus on the process rather than the product. Even though she postulates that controlled experimentation is the most valuable type of problem from the viewpoint of understanding science, she includes observation of the environment and demonstration of a basic principle as appropriate topics of study for an elementary science fair. Margaret McNay (page 54) also makes a convincing argument for the value of nonexperimental projects, pointing out that much of science is descriptive study rather than experimental.

How does one reconcile the differences of opinion in the literature? By taking a look at the goals of school science above, one can infer that an inquiry-based approach to science is needed; however, inquiry is not limited to experimentation. For example, volcanoes are okay as science fair projects as long as one has taken an inquiry approach to the study of the volcano. If the focus of the project is the making of the model, then a goal of science teaching has not been met. However, if the focus of the project is to understand volcanoes and their structure and to share that knowledge with others, which results in further inquiry and exploration, then the divine secret is understood.

Secret 6. Frustrating experiences can be positive learning experiences.

Linda Sittig and Cecelia Cope show how reflection can turn frustration into something positive. Sittig (page 36) shares her frustrations and reflections about her six-year-old daughters' science fair experiences from a parent's perspective. Cope (page 50) shares how she helps her "drained" middle school students reflect on their science fair experiences, resulting in a boost to classroom morale.

Secret 7. Implementing noncompetitive alternatives to the science fair, such as expos, carnivals, and festivals, can increase student, family, and community involvement.

Debbie Silver and others share this secret as they explore alternatives to science fairs. Silver (page 70), while working at a small, rural, elementary school, revitalized the science fair program by involving the community and including cooperative and noncompetitive options. In reading her article, one will note that she uses the scientific method to solve a problem.

- *Problem*: How can I increase the participation in the school's science fair?
- *Research*: Silver reviews the literature on science fairs and the impact of competition on participation. She also reviews various noncompetitive alternatives to science fairs.
- *Hypothesis*: If the competitive nature of the science fair is removed, then participation by students will increase.
- *Procedure*: Because Silver cannot decide which alternative to try, she implements all of the ideas and calls the event an expo. The events include the traditional, judged science fair, a share fair, class demonstrations, an invention convention, family physics fun festival, family science Olympiad, business exhibits, and special presentations from members of the community. She repeats the event over a multiyear period.
- *Results*: Even though participation in the expo is not required, most of the students in her school now choose to participate.
- *Conclusion*: Removing the competitive nature appears to have increased participation by students.

One may argue that other uncontrolled variables contributed to the success, such as community and family involvement; nevertheless, this small school has achieved something remarkable and solved its problem of low participation in a science event.

Daniel Wolfe (page 78), inspired by Silver's success, implements an expo for 4th through 12th grade students. He decides to keep the competitive component intact for the expo projects; however, he allows students to choose whether they want their projects judged.

Doug Cooper (page 82) shares his ideas for implementing a noncompetitive science event patterned after the traditional school carnival. He provides suggestions for organizing the event, and recommends setting up booths that explore the scientific phenomena associated with the games found at a typical carnival.

Similar in design is the science festival. According to Verilette Parker and Brian Gerber (page 86), the characteristic that distinguishes a science festival from a science fair is the interactive nature of the science exhibits that students share with each other. The authors offer several ideas for some popular interactive exhibits.

If you have experiences with science fairs or science fair alternatives that you would like to share with others, the editors of *Science and Children* and *Science Scope* want to hear from you. Please visit the NSTA website at *http://nsta.org/* for author guidelines.

References

Lowery, L. 1998. *The biological basis of thinking and learning*. Paper presented at a regional meeting of the National Science Teachers Association, Albuquerque, New Mexico (Dec.).

National Research Council. 1996. *National science education standards*. Washington, DC: National Academy Press.

About the author:

Donna Gail Shaw is a professor of elementary science education at the University of Alaska–Anchorage, Alaska.

Section I
Science Fairs

Preparation

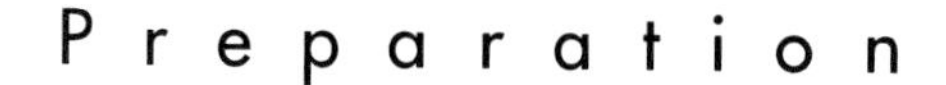

Getting a Jump on the Science Fair

Deborah C. Fort

The projects—116 of them in a public elementary school with 350 students—ranged from a test of the amount of dye in candy to a comparative study (by sex) of responses to breakdancing.

The principal, a former engineering and math major herself, required projects from the 50 graduating sixth graders But the success of the annual Murch Science Fair in Washington, D.C., had to do with much more than administrative mandates. The science teacher and the enrichment teacher, both veterans of 17 years of elementary school teaching—and of many science fairs—offered invaluable assistance to all interested contestants And several parent volunteers planned carefully, pitched in early, worked hard and faithfully, and stayed long hours.

Plan Ahead

If you're interested in avoiding last minute starts and finishes on science fair projects, if you want to witness the death of the artificial two-week (or even one-night) project, if you even dream—as the Murch principal does—of seeing a first-grade project extend into second and third grade and beyond, get started now.

Topic: science fair
Go to: *www.scilinks.org*
Code: SFP04

The first task of volunteers and staff alike was to help students struggle with the thorny problem of what, exactly, a science project is. Young children need some clear guidelines on scientific method, because the difference between a science and an art project can be something of a mystery to a kindergartner. Children in the lower grades can be taught to see science projects as opportunities for problem solving, for critical and analytical thinking, and for understanding cause and effect. Upper elementary students can sharpen research skills and discover new ways of conducting experiments.

Many educators believe that science fair projects should be part of the regular curriculum. Whether or not this is possible in your school, some fall preparation and continued work over the year before the big push in the spring will help you avoid some of the pitfalls that can diminish the value of science fairs or even make them harmful.

If a science fair project is an overall requirement, as it was for the Murch sixth graders, the science teacher—ideally in tandem with classroom instructors, volunteers, and the principal—should meet in the fall with all the children to give them an outline of scientific method, to

define the categories in which they might work, and to try to get them thinking about their special interests.

If the science fair is a voluntary affair, as it was for the younger Murch students, divide interested children by age (say, first through third graders and fourth through fifth), and begin small group discussions, like those offered by the enrichment teacher at Murch, as early in the year as possible. Because you will be approaching each child's project as a unique expression of his needs or her experiences, expect this procedure to take several meetings. Once the school staff has a sense of which children are interested in entering, provide volunteers with a list of their names. An early call to the parents of participants is likely to produce more help later, better sustained efforts on the parts of their own sons and daughters, and perhaps even some assistance for entrants who don't have strong support at home.

"I Can't Think of Anything"

The Murch science teacher finds that requiring her students each month to read part of any science magazine such as *Science News, Science Digest, Smithsonian, Ranger Rick, National Geographic World, Discover,* or *Scientific American* can help to focus students who think they have no ideas for projects. She couples weekly encouragement for all students with trips to the library as a counter to this kind of vagueness.

If a child expresses no specific interest but is tending toward something unfocused like "animals," encourage him to be more concrete. Does he want to know more about dogs, or does he want to learn about other animals? Does he have a dog? What kind? What exactly would he like to know about his uncle's Hungarian sheepdog that he doesn't know now?

If a child expresses no specific interest but is tending toward something unfocused like "animals," encourage him to be more concrete.

Know Thyself

The best science fair projects grow out of something important in the child's life. For example, one Murch eight year old, the owner of a mixed breed collie and German shepherd, studied the responses of various neighborhood dogs to recorded wolf howls. A second grader who had suffered a stroke as an infant studied his own ability to exert self-control through biofeedback.

Science fair projects should be in progress certainly by October, ideally by summer or even earlier. They should stimulate children to more ambitious future projects. In addition, most projects should involve mathematical skills; they should encourage organizational ability; and they should show children how to demonstrate method and results. However, not all science fair projects need be experiments calling for hypotheses and conclusions. Children can also learn a great deal from nonexperimental projects like those suggested by Margaret McNay on pages 54–56 of this book.

According to the Murch principal, what begins as a part-time effort should eventually be refined or expanded so that a project started one year can

Ideally, judges talk with each child and respond to oral input individually, allowing presenters to be proud and informative.

carry over into another year—and beyond. Teachers can encourage the choice of and commitment to longitudinal studies in topics like human growth and development (children can study themselves), methods of energy saving, changing purification systems, and conservation projects (what begins as an experiment involving plants can become part of a school garden).

"The Matthew Effect"

Like a strong science program, a science fair must serve all students—not just those lucky enough to come from homes where science is valued and practiced. Otherwise, we run the risk of further hurting children who already suffer from disadvantaged home environments—of intensifying what R.K. Merton has called, "The Matthew Effect." The words of the parable—"whosoever hath, to him shall be given, and he shall have more abundance, but whosoever hath not, from him shall be taken away even that he hath" (Matthew 13:12)—whatever their meaning in Christian theology, should not be an apt description for our children, particularly those who are not lucky enough to have parents who support them academically. Such children, more than the offspring of the privileged, need superior teachers as well as contact with peer academic achievers.

The Biggest Show in Town

In the best of all possible worlds, each elementary school student would do a year-long project. The Murch science teacher believes that she should help all children who want to enter rather than helping only a few intensively. Although the latter method does seem to produce winners, it is often hard to tell where the child stopped and the professional started. In addition, this approach focuses too much on winning and can deprive the majority of children of the chance to participate (or at least to do so without expert help). Parents and other volunteers *can* offer valuable help, but understanding the appropriate limits of that assistance is very important. Written explanations of who did what, like that offered by the grandfather of one young solar engineer, offer a possible solution to an old dilemma.

If you agree that the biggest science fair is the best, you should think about logistics now. Line up parent volunteers this month, for example, unless you were so well organized as to have asked for them on the first volunteer sign-up sheets sent out last month. And, if you're in a school with many students and not much display space, maybe you'll want two science fairs—one for the lower grades and another for the upper ones. Careful planning now can lead to a fine show in the spring science fair.

Labors of Love

The parent organizers of the Murch fair included a child psychiatrist, an artist, and a U.S. government administrator, whose hours of help

(continued on p. 9)

Wilbur and Orville Started Out on the Ground

A judge at the Murch Fair commented on her search for creativity, imagination, and scientific method, which she defined as "a question leading to an answer (though not necessarily the one expected)."

One fifth grader's winning study of "Heart Music" fitted her definition. He attempted to measure the impact on the heart of different types of music as measured by an electrocardiogram (EKG), loaned by an obliging uncle who worked as a doctor at a local hospital. His subjects were his 34-year-old uncle (76.5 kg), his 30-year-old aunt (51.5 kg), and his 8-year-old brother (22.5 kg). After learning how to use the EKG machine from his uncle (who also assisted by hooking up the participants), the investigator first tested his subjects' hearts without music, then as they listened to "Footloose" ("a fast rock and roll song"), a Mozart concerto (a "slow, calm, classical piece"), Frank Sinatra's rendition of "New York, New York" (a "medium-paced song with a strong beat"), and the Beatles' "I Want to Hold Your Hand" (a "calm rock and roll song").

The young scientist's hypothesis, the faster the music, the faster the heart will beat (and vice versa), was *not* supported by his data.

He concluded that he had ignored many variables. For instance, he realized that the readings could have been affected by what the subjects had eaten, by their physical condition, by the volume of the music, and by their emotions. "For example, my brother laughed the whole time he was tested," a fact that perhaps contributed to his wild EKG.

Science Triumphant

Some of the hypotheses at the Murch Fair were proved.

One third grader studied hydrilla, which he described as "a noxious weed taking over the Potomac" [river] and which his display asserted was "wanted dead or alive." Early in October, he and his family went down to a dock in the Potomac to collect hydrilla samples. He put 0.070 g of hydrilla in each of 26 large Mason jars also containing various concentrations of river water, salt, and mud. The experiment stank so foully that his mother removed it to the attic at Thanksgiving where it eventually cleared and stopped stinking. Over the Christmas holidays, the hydrilla seemed to die and decompose; however, when the experimenter brought it back downstairs in January, almost all of it came greenly and slimily back to life.

The hydrilla died permanently only in very salty concentrations; mostly its decomposition was followed by a total regeneration accompanied by tiny snails whose eggs must have been in the samples the student collected in the fall.

Washingtonians, especially ones who enjoy boating on the Potomac, are deeply concerned about the spread of hydrilla. They wonder how far upriver it will spread—this Murch scientist thinks he knows the answer. Look for the snails, and you'll later find the weed.

Last But Not Least

Inspired by Laura Ingalls Wilder's *Farmer Boy*, a second grader experimented with the best ways to preserve ice. According to Wilder's title character, ice blocks 20 inches square cut from frozen lakes and buried in sawdust "would not melt in the hottest summer weather. One at a time they would be dug out, and Mother would make ice-cream and lemonade and cold egg-nog."

The investigator measured the room temperature, gathered his materials, surrounded ice with various substances, and recorded the time when the ice first began to melt. He studied the insulating ability of aluminum, water, paper, dirt, sand, and sawdust. Like Wilder's farmers, he found that sawdust works best.

Finally, an artistic as well as scientific upperclassman studied "Breakin." His full-color exhibit featured a large illustration—a brown man twirling on one hand (gloved in fingerless mitts), blue pants and red sneakers reaching for the ceiling. The scientist was trying to discover whether boys would react to break-dancing more than girls, and his hypothesis was that males would be more responsive. He observed, music box in hand, on the playground for several days, finding "Few girls stopped playing, but most boys stopped playing and began to do [move] to the rhythm of the music. … There were 11 boys who did a complete routine."

(continued from p. 6)

provided far more than elbow grease, scissors, paste, and lunch from a fast food restaurant on judgment day (though their contribution included these items). Concerned that the children not become confused that science fairs "are" science and that the fairs offer genuine opportunities for learning, not just ones for last minute competitiveness, the parents combined their offers to carry and to set up heavy and complex projects with an equally important willingness to discuss and, at times, explain concepts to the children.

Parental helpers stressed the need for good lines of communication among faculty, administrators, and volunteers regarding practical matters like time tables, resources for assisting students with project assembly, and availability of materials. They planned a November letter to parents signed by the principal and science teacher as well as by them to explain the philosophy of the fair and to encourage participation. They backed this overture with approaches to the students in the once-a-week science classes and in follow-up calls to parents of interested children.

After working on projects at home over the winter holidays and into January, students were encouraged in mid-February to submit written summaries of their work to the science teacher for her input. The principal made the science teacher available to participants in each class one hour a week; the sixth graders, whose participation was required, got more specialized attention. Then, during the week before the fair, parental volunteers were available after school to help with lettering, pasting, and assembling, as well as encouraging and explaining.

Set-up day was the Saturday morning before the fair (the weekend hours guaranteed a good deal of volunteer help from working parents). After the projects were judged on Monday, every class had a chance to walk through the exhibit hall and inspect the projects. The principal also arranged for early morning and late afternoon hours, so that parents and other interested spectators who worked would be able to see the science fair exhibits.

...the most important contributors to the success of the fair were the children—their commitment, their time, their imagination.

But the most important contributors to the success of the fair were the children—their commitment, their time, their imagination. About a third of the students offered projects in the following categories: behavioral science (23 projects), biology (25), botany (7), chemistry (6), mathematics and computers (2), physics (28). As they set up their exhibits and stood beside their completed projects, the children were glad to offer comments to the classmates and parents who came to ask questions and admire.

Here Come the Judges

Picking the judges—how many and with what qualifications—is a decision that is important to the success of your school's fair. If the judges are part of your school's community, their anonymity should be closely guarded. This precaution

is, of course, unnecessary if the judges are chosen from outside.

When picking judges—and you should get as many as possible, preferably enough that each project can receive several evaluations before the results are averaged—try to find flexible scientists and educators who will be willing to respond to the projects as wholes and who will not lose sight of the creativity that may fuel an imperfectly presented project. Before the judging begins, present each judge with a set of the criteria students have followed in creating their projects.

Whatever your particular criteria, the judges should note the display and, as relevant, the hypothesis, method, data collection, and conclusion, as well as the level of understanding the student demonstrates through the display and in response to questions. Ideally, the judges should be able to talk with each child and respond to his or her oral input individually, offering each presenter time to be proud as well as informative.

Murch's judges, recruited by the science teacher and the parent volunteers, included a chemist from a local hospital, an educator from a neighboring state's public school system, a military scientist, and a junior high science teacher.

There Go the Judges

Another judging possibility—albeit a heretical one—is to give all participants A's. At one recent fair, the only acceptable judgments were Superior, Outstanding, and Noteworthy. In any case, make sure that all contestants win something—a ribbon, a certificate, or a medal.

Or, even more of a violation of the American spirit of competition, dispense with judgment altogether. Making each child's science fair project part of the regular science curriculum would render public ratings unnecessary, and everyone—from the most advantaged student to the least—would have a chance to participate. Such a procedure would also help to separate the parental contributions from the children's.

Fair Enough?

Once the projects have gone back to homes and (unfortunately) sometimes to trash cans, try to keep the memory alive to fuel enthusiasm for next year's fair and for this year's achievements.

For instance, do a follow-up unit on a particularly impressive project. The Murch enrichment teacher had her third graders create a book called *Hydrilla Monster* based on one child's project.

In addition, she asked some significant questions:

- Did you like your project? (yes or no)
- Did you find out everything you wanted to know about it? (yes or no)
- Would you like to continue learning about it? (yes or no)

Although the results were mixed on the first two questions, a resounding 90 percent responded affirmatively to the third question.

So, if your students react similarly, note that fact and encourage them to get started early next year on science fair projects that are logical extensions of the experiments they seem unwilling to abandon.

A science fair wrap-up is also a good occasion for catching the attention of those students who didn't participate this year and who may wish now they had. Encourage them to start thinking now about areas that could become a project for the future. Help them to picture themselves as part of the fair next year even though they missed out this time.

Mastering the Science Fair

Ruth Bombaugh

Do you feel overwhelmed by details at the very thought of a science fair? You don't have to. Use this master schedule (pages 14–15) as your checklist, and spread those tasks out over a year's time.

I've been perfecting this schedule for about ten years, so I know it works. I began to develop it because as a young teacher I was a lot like my seventh-grade students—long on energy and enthusiasm, but short on organizational skills. The details became unmanageable.

Even a seemingly innocuous detail like arranging space for the fair can cause big problems if it's not attended to far enough in advance. The first science fair I organized was elbow to elbow with 180 students packed into the cafeteria. Now I reserve the gym a full year ahead of time.

I start by going over the basketball schedule with the athletic director to make sure the gym is free, and my seventh graders aren't scheduled for an away game on the night I want. Next I explain to the gym teacher that we'll need to set up on the day of the fair, and offer to trade spaces with her for that one day. Finally, when I've gained the cooperation of both the athletic director and the physical education teacher, I go to the principal and make a building request in writing. I also reserve the cafeteria as a hospitality area for the evening of the fair. This gives parents a place to be while the judging is going on.

Preparations inside the classroom also begin a year in advance. Each spring, I visit the various sixth-grade classes to tell them about the science fair. This gives the students lots of time to start thinking about science projects. I bring some of my most successful seventh graders along to demonstrate their projects. I share our "brag book" of pictures, newspaper clippings, and other evidence of the recognition my students have won at district and state fairs. This introduction to the science fair stirs up enthusiasm and anticipation.

The next fall, I give the students experiences with hands-on lab work. They learn the scientific method by performing controlled experiments. The students practice for the fair by writing up several of their experiments as formal summaries, including question, hypothesis, materials, procedures, results in a chart or graph form, and a conclusion.

Early fall is also the time to draft a schedule specifying the minimal requirements for the science fair and setting a due date for each

requirement. The six requirements are: (1) performance of an experiment with data collection, (2) a formal summary of the experiment, (3) a research report with bibliography, (4) a visual backdrop, (5) an oral presentation, and (6) attendance on the night of the fair.

Structure is vital to the success of the fair. The schedule of steps and due dates provides the solid framework middle school students need. A number of my "learning disabled" students have won high honors at district fairs thanks to the structure the ten-week student schedule gave them.

Communication between home and school is another vital element to planning a science fair. At the first parent/teacher conference, I give parents a copy of the schedule and a letter that fills in the details. Parents are consistently supportive when they know what will be expected of their children.

Students spend three of the total ten weeks deciding on a project. Choosing the right project is the most essential step of the whole process. I don't want any student to work on a project that is so undemanding he or she won't learn from it, but I also don't want projects so difficult that students are set up for failure. Most seventh graders have never had to make choices of this kind before, so they need patient guidance.

To be fair to the students, judging criteria are based on the requirements they have been asked to meet. I give copies of the judging sheets to students well in advance so they know how their projects will be rated. They will earn 45 of the 100 possible points just for meeting the basic requirements. Knowing this motivates them to keep on schedule. The remaining 55 are quality points which indicate how well they meet the requirements.

When science fair day finally arrives, I can relax and enjoy myself. It's like a wonderful party! Students are dressed up and on their best behavior. The gymnasium has a holiday air. I make sure that I am free to greet the judges, who are local professional people. I put a high premium on student/scientist interaction. I don't assign judges more than four to six projects each. Students spend a class period after the fair writing personal thank-you notes. As a result, judges are eager to keep coming back year after year.

After the school fair, the students designated to go to the district fair get together to further improve their projects. Again, structure and communication are essential, so I arrange a time after school for each student and a parent to meet with me, read the judges' comments, and draw up a contract for the tasks the student agrees to do as part of our school's science team. These tasks include preparing the oral presentation for videotaping and practicing it in front of the sixth-grade classes when I make my spring visits. This completes the yearly cycle that began the previous spring.

Structure is vital to the success of the fair. The schedule of steps and due dates provides the solid framework middle school students need.

Master Schedule for

During the School Year Previous to the Fair

Preparation Outside the Classroom

Reserve the gymnasium for the whole day of your fair. (Talk to athletic director, gym teacher, principal.)

Reserve the cafeteria as a hospitality center for the evening of your fair.

Urge fellow teachers to assign your prospective students a research report.

Plan the format of your fair with the other teachers who teach the same grade. Possible options include:

1. An Interdisciplinary Science Fair: all students do a science project but the library research is a social studies assignment, the backdrop is an art assignment, the graphs are a math assignment, etc.
2. An Academic Fair: all the teachers cooperate, and students may choose to do either a math, social studies, science, or language arts project.
3. A Science Fair: only the science teacher oversees.

Preparation Inside the Classroom

Visit the science classes to tell all prospective students about the science fair.

Schedule the current science team to give presentations to the prospective students and display their finished products.

Help any interested students to design science fair projects which they can work on during the summer.

Before Students Start to Work on Their Science Fair Projects

Preparation Outside the Classroom

Prepare letter for parents which states requirements that students must meet.

Prepare week-by-week schedule for students telling what they should be working on and what deadlines they should meet.

Prepare judging sheets.

Prepare award certificates and order ribbons. (Any student calligraphers?)

Reserve public library and school library time for students to be shown reference materials. Arrange to have reference materials in the classroom too.

Reserve space and time for awards assembly.

Preparation Inside the Classroom

Stress hands-on lab experiences with data collection in your science classes. This reinforces concepts and helps students learn the scientific method in a concrete manner.

Require students to write up their lab experiments in science fair form. Make sure they have all the parts of an experimental summary—question, hypothesis, materials, procedures, results in a chart or graph form, and a conclusion.

Director of the Fair

During Student Preparation of the Science Fair

Preparation Outside the Classroom

Contact resource people when they are needed for assistance.

Two to three weeks before the fair, line up your judges. (Personal contact by telephone works best.)

Preparation Inside the Classroom

Give it your all!

Follow the week-by-week schedule, and anticipate students' need to learn new skills. Teach how to write bibliographies about a week before they're due.

In the Ten Days Leading Up to the Fair

Preparation Outside the Classroom

Arrange hospitality. (Your school secretary, principal, and/or fellow teachers may be willing to be hosts. Could the home economics students bake cookies?)

Make up the judging assignments and group sheets for each judge.

Make up name tags for the judges.

Arrange to have a volunteer photographer.

Alert the media (newspapers, radio, local TV).

Set up the tables and a microphone the evening before the fair.

Preparation Inside the Classroom

Continue to follow the week-by-week schedule and DON'T PANIC. The last-minute Lizzies will often do wonders when the time crunch is on!

Go over the judging sheets in class and have students fill out the tops: name, date, title, number of project.

On the Actual Day of the Fair

During the School Day

Have students set up their projects during the class periods.

Let other grades view the projects, with your students serving as hosts.

Remind students to dress well for the fair and to be polite.

During the Fair in the Evening

Be sure you have delegated as much responsibility as possible. This involves more people in the fair and leaves you free to trouble-shoot.

Greet your judges! (They are V.I.P.'s.)

Enjoy yourself!

After the Fair

Finishing Up Local Fair

Average the judges' scores.

Fill in names on award certificates and host awards assembly.

Write articles for the newspapers.

Have students write thank-you notes to judges.

Preparation for District Fair

Draw together a science team of student volunteers and meet with each parent and student to draw up a contract of responsibilities.

Help each student follow through on his/her contract.

Videotape the science team.

Prepare for Science Fair

Susan Shaffer

Looking for a unit that will quickly get students on task and allow them to use science process skills and the Internet? Preparing your students for a science fair project will accomplish all of the above. I use computers and the Internet to help students prepare, research, carry out, and present their projects.

Students work in pairs at five classroom computers with online connections. In preparing students to do their projects, I introduce them to a web browser program preset to open up to a homepage I designed with Claris Homepage. This homepage displays links that I have found useful in teaching the science fair process (see Resources).

Because this is typically the first opportunity my students have at using a web browser, I take a very structured approach. I prepare a corresponding worksheet, How to Do a Science Fair Project (see page 17, top), that guides students step by step to open the web browser, click on the links, read the material, view the photographs, and answer questions at each of the websites. In this way, students learn how to use the software while also gathering information on doing science fair projects. One-third of the class works on the Internet, another third watches the video *How to Prepare a Science Fair Project* (see Reference), and the remaining third works in pairs on a worksheet I designed to correlate with the handbook accompanying the video. As students rotate through the computer, video, and handbook stations, they learn through three different approaches.

Next, I assign a second worksheet on science fair topics and ideas that corresponds to different links on the same homepage. These sites highlight completed projects and offer lists of topic questions. While some students work on the computers, others look for project ideas in classroom books and resources. After completing the online overview of topics and the hard-copy search, students generate a list of possible science fair topics.

Students also become proficient in conducting library research on their chosen topics. They use InfoTrac to look through periodical listings, Catalog Plus to search for books in local libraries, and CD-ROM encyclopedias to gather information. Once students have gathered their research materials, they use word processing programs to write literature reviews and research papers on their topics. After students

Worksheet for a Science Fair Project—One Teacher's Approach

How to Do a Science Fair Project

Start by double-clicking on the web browser program icon. A homepage titled "Science Fair" will appear onscreen. Find the link "Steps to Prepare a Science Fair Project" and click on it. Read each step and then answer the questions below.

1. How many steps are there to doing a science fair project?
2. List the steps and write a sentence describing each one.

Click on the "Back" button to return to the homepage. Find the link "Practical Hints" and click on it. Read the hints and answer the questions below:

3. How many hints are listed?
4. List any four of the hints on your sheet of paper.

Click on the "Back" button. Find and click on the link "Outstanding Projects." Click on the link "1996." Find the project done by Julie Burris and click on the word "Click" near her project.

5. What happened when you clicked on "Click"?
6. List all of the titles you see on her backboard.
7. What is her question?
8. What is her hypothesis?
9. Does her conclusion agree with her hypothesis?

Click on the "Back" button to return to the homepage.

Presentation

Click on the link "Science Fair Workshop." Read the text and answer the questions below.

1. What type of first impression do you want to make on science fair judges or viewers?
2. What four things should your display do?
3. How can you use colors to your advantage on your backboard?
4. What should the project title do?
5. List the information that should be on each panel of a three-sided exhibit. Use the diagram below.
6. List two ways to make your display look professional.
7. List two problems you might run into.
8. List three things you should do during your oral presentation.
9. List all items that your written report should contain.

complete their experiments, they return to the computers again, using a spreadsheet program to organize, graph, and analyze the raw data.

After the topics have been chosen, the library research completed, and the experiments done, students use my homepage one more time with a worksheet titled Presentation (see page 17, bottom). Again, one-third of the students work on the worksheet at the computers, another third uses the handbook to answer worksheet questions, and the remaining students view projects from last year, on display in the classroom. Many of the previous year's science fair winners are invited to class to give advice on the presentation and procedures in science fairs.

I have found that using computers motivates my students to produce high-quality research papers, backboards, and presentations. Some of the students have even won at the county-level science fair. Our next challenge is publishing these winning projects on the Internet.

Reference

How to prepare a science fair project. 1993. 26-minute videotape. Catalog #10690VA, $95. United Learning, (800) 424-0362.

Resources

- Steps to Prepare a Science Fair Project: *www.isd77.k12.mn.us/resources/cf/steps.html*
- Science Fair Homepage: *www.stemnet.nf.ca/~jbarron/scifair.html*
- Science Fair Workshop: *www.eduzone.com/tips/science/showtip4.htm*
- Topic sheets: *www.eduzone.com/tips/science/second.htm*
- Project resource guide: *www.ipl.org/youth/projectguide/*
- Practical hints: *www.scri.fsu.edu/~dennisl/CMS/special/sf_hints.html*
- Outstanding projects: *www.oxnardsd.org/campus/frem/sci/sfp.html*

Project Selection

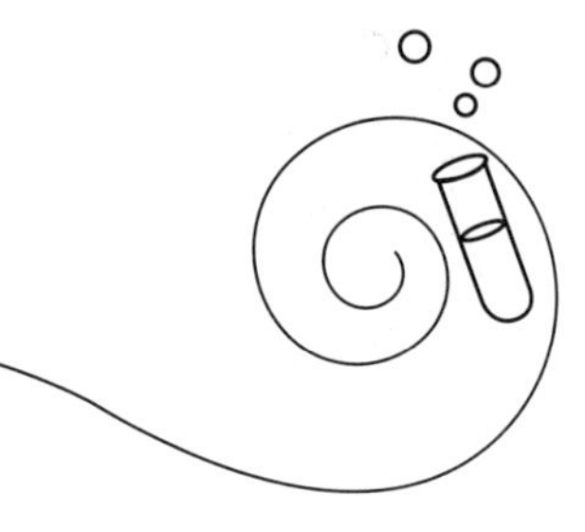

"Oh No! A Science Project!"

Gail C. Foster

Mention the words "science project" to a teacher, student, or parent, and you'll probably provoke reactions ranging from delight to aversion. More than one elementary science teacher has been confronted with wails of "Oh, no, a project!" when the assignment was introduced. And more than one (if he or she is honest) will admit to harboring occasional doubts about the value of doing science projects.

The problem may be that people involved in working with science projects sometimes forget what the projects are supposed to accomplish. The primary purpose is to encourage students to think critically and to investigate. A successful project has given its creator a chance to observe, infer, measure, identify, classify, hypothesize, experiment, manipulate variables, and interpret data. It has helped the student learn how to learn.

Topic: inventors
Go to: *www.scilinks.org*
Code: SFP20

Guess What?

Science projects often cause difficulty because they appear out of nowhere, like a rabbit out of a hat. And children are supposed to be able to do a project just because the school is having a fair or because the teacher says that a certain percentage of their science grade depends on it. It's not that easy. Children may not have the process skills needed to do such a project. They lack extensive practice in observing and making inferences, and they may not even have participated in any simple group projects. Requiring a child to do an individual project without this experience is like introducing the alphabet and then expecting the child to write a novel.

Just Pick a Topic

Selecting a topic and identifying a problem are undoubtedly the most difficult parts of doing a science project. Common approaches include having children write down several "areas of interest" or pick topics from a list of time-honored favorites. If these methods don't work, children may be sent off to the library or media center, where they find "cookbook" experiments and then organize them into the form the teacher has specified. At best these approaches are artificial; at worst they cause children to work on topics in which they have no real interest.

A more productive approach is to introduce students—early in the school year—to the idea of asking questions about the world around

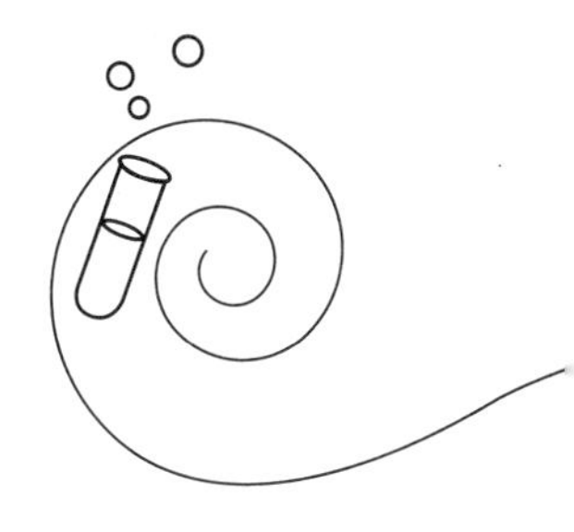

them. Asking questions shouldn't be too hard, should it? Consider children's natural curiosity: "Why do balls bounce?" "What makes a radio work?" "Why are leaves green?" "What do spiders eat?" Unfortunately, as Lazer Goldberg (1979) notes, children seldom ask such questions at school, so you'll need to get them going. Set the stage for questioning by taking your class outside or gathering them around an aquarium, an insect zoo, or a learning center. A bulletin board with a new display could serve as a focus for questions. Talk about things you might have wondered about. Do ant lions turn into anything else? Will spiders eat dead insects? Does salt water boil as rapidly as fresh water? How does soap clean things? Will certain fish react to seeing themselves in a mirror?

Children respond quickly to this approach and begin contributing questions of their own. Explain that some questions can be answered by investigating while others cannot, and give some examples of both. Select some questions and have the children tell how they could find the answers. If time permits, groups could design simple investigations to answer simple questions. And be prepared for additional questions that arise from observation. For example, if children have experimented and found that they can lower the temperature of water by adding baking soda, they may have new questions that need answering: Will the water freeze if they continue to add baking soda? Will the temperature of hot water drop the same number of degrees as that of cold water? Will adding baking powder lower the temperature? How about yeast?

Creating Excitement

Once students are accustomed to posing their own questions, you can set the stage for selecting a topic and identifying a problem using the same approach you employed earlier to stimulate students' curiosity.

If students seem less than enthusiastic, blitz them with attention-capturing activities: try optical illusions, "eyeball benders," mystery boxes,...

First, assess your class's attitude toward science projects. If students seem less than enthusiastic, blitz them with attention-capturing activities: try optical illusions, "eyeball benders," mystery boxes, mixtures and fluids, puzzles, tangrams, and magic tricks. Contests that pose problems with many possible solutions can be particularly stimulating. (Who can figure out the best method for floating an egg?) You can find suggestions for such contests or activities in Joe Abruscato and Jack Hassard's *The Whole Cosmos Catalog of Science Activities* (Santa Monica, Calif.: Goodyear Publishing Co., 1977).

You can also arouse curiosity by having students contribute to a classroom resource center. This resource renter might include

- a mini-museum of interesting items such as shells, galls, lichens, seeds, magnets, magnifying glasses, a thermometer, batteries, wire, bulbs, a stethoscope, a tuning fork, balloons, candles, a funnel, a compass, pulleys, balls of various sizes, a gyroscope, a prism, a paper airplane, convex and concave lenses, marbles, a stopwatch
- a bulletin board with a word-and-picture col-

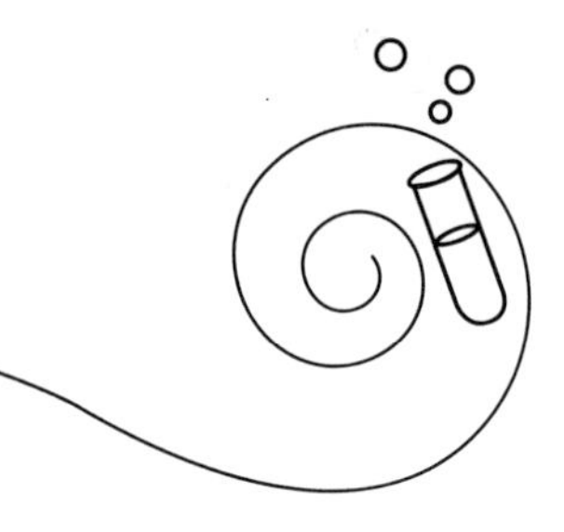

lage depicting subjects that might spark ideas

- a display of advertisements for competing products and the containers the products come in to encourage comparisons
- interesting books and magazines like *Insects, Optical Illusions, Ranger Rick's Nature Magazine, National Geo World, 3-2-7 Contact*
- lists of familiar places, people, and things that might inspire your students' investigations:

  personal interests—hobbies, pets, or leisure-time activities

  home—under the sink; in the refrigerator, pantry, garage, or yard

  neighborhood—school grounds, parks, vacant lots

  stores—grocery stores, drugstores, bookstores, malls

  people—family, friends, classmates

  media—television, movies, radio, newspapers, books, magazines

Help for the Perplexed

In spite of all this, some children will need additional suggestions. Help them arrive at topics

Trying Things Out

With a little practice, your students can become accustomed to locating topics, posting questions, and formulating hypotheses. Discuss the first four examples below as a class. Then, have students work together to fill in the blank columns in the last six items.

Topic	Problem Statement	Hypothesis
1. Conserving water	Do showers take less water than tub baths?	Showers take more water than tub baths.
2. Hexing things	Do thick liquids boil as fast as thin liquids?	The thinner the liquid, the faster it will boil.
3. Balls that bounce	Will more air make a basketball bounce higher?	If air is added to a basketball, then it will bounce higher.
4. Ant lions	Do ant lions only eat ants?	If a worm or small insect other than an ant dropped into an ant lion's hole, the ant lion will eat it.
5. Vision	Does eye color have an effect an pupil dilation?	
6. Rusting		Iron nails rust more quickly in salt water than in fresh water.
7. Soil	Does water soak into some kinds of soil faster than others?	
8. Plants		
9. Solar heat and color		
10. Hearing		

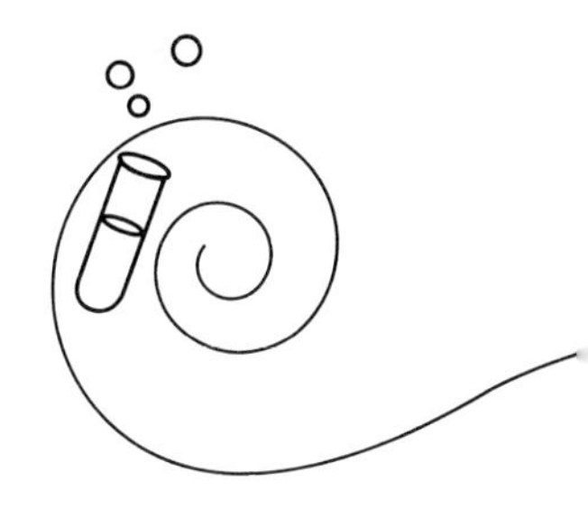

by staging brainstorming sessions, where class members contribute ideas and post the results for pondering. Or organize a Student Advisory Committee for Perplexed Individuals.

If a few children are *still* without a problem to investigate, you can fall back on teacher-child conferences, lists of specific projects, or a collection of experiments.

What Now?

When a child is ready to state the problem for the science project, he or she needs a brief conference with the teacher. In most cases you only need to help the child narrow the problem to a question that can be answered by experimentation. Let the child tell you what he wants to find out. You may need to help him simplify or reword the problem, but resist the impulse to tell the child what he wants to know. Simply ask: "What do you want to know?" and "How will you find out?"

Before beginning an experiment, the child should be able to express exactly what he is trying to find out (the problem) and how he plans to find out (the method of investigation). Theoretically, the child should find out as much as possible about the problem through observation and research. However, if the child actually researched the problem thoroughly, there would be few investigations because most answers to simple questions can be found in books. For an elementary school child, then, research should be highly specific and brief. (One side of one page will suffice for most projects.) Sources may include audiovisual materials, interviews, and brochures, as well as encyclopedias and magazines.

Table Talk

When your students are about ready to settle on their topics, use the chart on the previous page to give them practice in narrowing broad topics to specific problem statements and verifiable hypotheses. Work through the first four topics on the chart with the whole class. After each problem statement, ask students to design a simple experiment that would answer the question.

By the time you have presented the fifth topic, students should be able to proceed with the rest of the chart individually or in pairs. After students have completed the chart, present them with a blank chart and ask them to fill in the whole thing. (They might do this individually, in pairs, or as a group.) Stress simplification. The simpler (and therefore more manageable) the problem statement and hypothesis, the less difficulty the student will have completing the project.

A successful science project should begin with wonder and foster wonder. As Cornell professor Verne Rockcastle says, "After all, science is not a subject; it is a way of looking at the world around us. Good science teaching helps children develop ways of finding out what makes things happen, and what will happen if... ."

References

Goldberg, Lazer. "I Know the Answer, But What's the Question? *Science and Children* 11:8-11, February 1979.

Rockcastle, Verne N. *Some Basic Philosophy of Good Science Teaching in Elementary Schools.* Atlanta: Addison-Wesley Publishing Company, n.d.

Resources

Cornell, Elizabeth A. "Science Fair Projects: Teaching Science or Something Else." *Current: The Journal of Marine Education* 3:17-19, Fall 1981.

Shephardson, Richard B. "Simple Inquiry Games." *Science and Children* 15:34-36, October 1977.

Smith, Norman F. "Why Science Fairs Don't Exhibit the Goals of Science Teaching." *The Science Teacher* 47:22-24, January 1980.

Ukens, Leon. "Inquiry with Toys." *Science and Children* 15:20, October 1977.

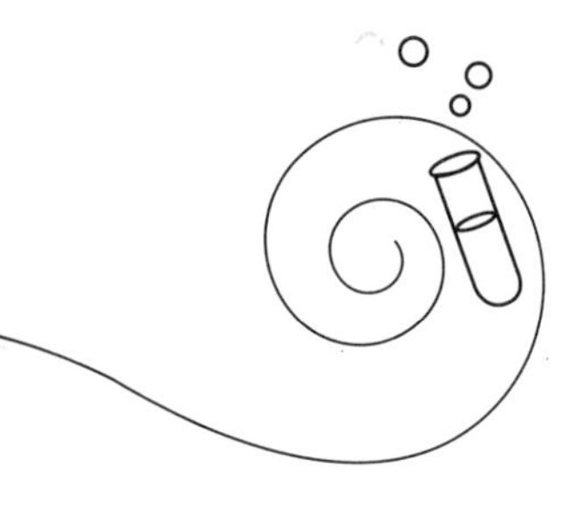

Scientific Investigations

Stephen C. Blume

The fundamental purpose of children's science fair projects is to develop critical thinking that can be applied not only to science but also to other subject areas including, ultimately, reality. The best projects stretch the students' investigative skills furthest.

The elementary school science teacher can foster children's exploratory techniques by training them to use the scientific method in the creation of their science fair projects. Although this method asks that investigators (1) state problems, (2) list required materials, (3) outline procedures, (4) formulate hypotheses, (5) record observations, (6) analyze data, and (7) draw conclusions, elementary school teachers can simplify the process by having young scientists ask these six questions:

- What do I want to find out?
- What materials do I need?
- What should I do with the materials?
- What should happen?
- What did happen?
- Did I find out what I wanted to know?

Following these steps, students formulate questions their projects will help answer, perform investigations, and collect and analyze data to arrive at a conclusion or new understandings.

Lists of science project topics or problem statements of experiments are of limited assistance to science teachers. Resources are as near as the indexes of science texts and instructors' guides, to say nothing of the comprehensive creativity of children. However, problem statement frameworks can help the science teacher and the students begin. For instance, students can move from problem to hypothesis following the suggestions in the box on the next page.

Projects that involve students in critical thinking and science process skills are best. I believe that show-and-tell displays of hobbies or models, laboratory demonstrations from science textbooks, or report-and-poster displays based on scientific literature are usually less valuable learning experiences because such efforts don't reflect science teaching's primary goals—developing critical thinking and investigative skills.

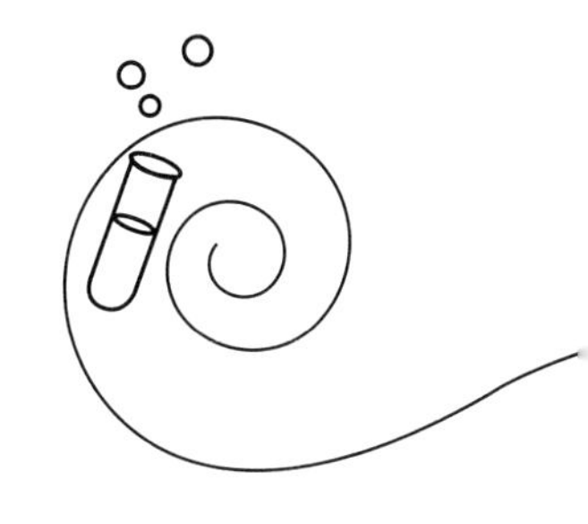

Moving from Problem to Hypothesis

1. What is the effect of ____________________ on ____________________?

detergent	germination of seeds
eye color	pupil dilation
light	growth of plants
temperature	the volume of air
oil	the growth of beans

2. How/to what extent does the ____________________ affect ____________________?

length of a vibrating body of light	sound
color of light	the growth of plants
humidity	the growth of fungi
color of a material	its absorption of heat
viscosity of a liquid	its boiling point

3. Which/what ____________________ [verb] ____________________?

paper towel	is	most absorbent
foods	do	mealworms prefer
detergent	makes	the most bubbles
paper towel	is	strongest
peanut butter	tastes	the best

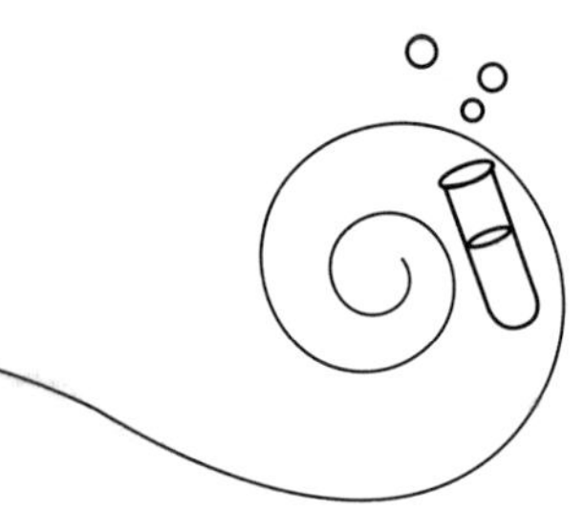

The One-Hour Science Fair

Charlene M. Czerniak

Getting middle-level students to conduct scientific investigations is often a difficult task. Children frequently choose favorite topics that do not lend themselves well to scientific investigation, such as How Did Dinosaurs Die? The Solar System, and Volcanoes. After teaching elementary and middle school science for ten years and judging in numerous science fairs, I have discovered that a child can be taught the processes and investigative skills for completing a science fair project in just one hour using chocolate chip cookies or paper towels.

These and other household materials can be used to teach students both quantitative and qualitative research approaches whereby students design an investigation and test variables related to simple, easy-to-understand topics. Students learn the steps involved in proposing a question, stating a hypothesis, designing an investigation, graphing data, forming conclusions, and making a display in a very short time period—all within an hour! This experience creates a wonderful beginning to get students involved in designing and completing a more in-depth science fair project.

A Quantitative Project

An easy way to introduce students to quantitative research is to pose a simple question that can be answered with quantitative data. For example, purchase four or five different brands of paper towels and ask your class the question "Which brand of paper towel is the best?" After allowing students a few moments to give their opinions or predictions, guide them to think about the variables that make a good paper towel. Some possible variables are thickness, price, absorbency, durability, how easily the paper towel tears away from the next towel, and how much lint it leaves behind.

Next, tell students that they will test each of these variables. It is important to let students design their own tests. In doing so they are forced to use problem solving and thinking skills to devise a good test. To keep the project within the one-hour time frame, stipulate that the test must be carried out within the classroom with materials from the classroom or science supply room.

After students complete their tests or investigations, draw a chart like Table 1 on the blackboard or on a transparency. Have students rank their brands of paper towel from one to five (best to worst) for each variable.

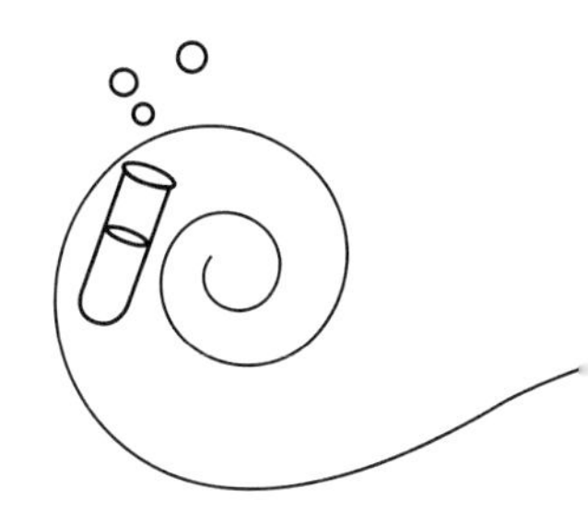

Lead the class in a discussion of how the data on the chart can be used to answer the original question, "Which brand of paper towel is the best?" Finally, guide the students to see how this information could be displayed for a science fair. Have students state the question, the procedure, the hypothesis, the results of the tests, and the conclusions. Draw a picture on the board of what the display might look like with the title, abstract, data chart, and other explanatory information.

Other simple quantitative science topics that children could investigate include

- Which brand of popcorn pops fastest?
- Which brand of diaper holds the most liquid without leaking?
- What shape paper airplane will fly the farthest?
- Which type of size C dry cell lasts longer, regular or alkaline?

A Qualitative Project

Students can also learn the processes involved in a scientific investigation using more qualitative measures. For example, ask your class the question, "Which brand of chocolate chip cookie is the most popular?"

Students will come up with variables such as softness, taste, smell, appearance, size, thickness, number of chocolate chips per cookie, and number of cookies per package. The quantitative measures can be handled in the same manner as described earlier. However, the more qualitative variables pose a new dilemma. Taste, smell, and appearance are matters of opinion. Lead students to understand that these types of variables are often tested by asking many people their opinions. Discuss popular television ads that purport claims such as "Most people prefer Brand X Cola" or "Nine out of ten people chose Brand X in a blind taste test."

Have students decide how they will test for taste, smell, and appearance. Help them determine a good number of people to test.

After students devise and complete their tests, have them make conclusions in the same manner as described earlier. Then have students draw a sample display board to explain the results. Other simple qualitative science projects include

- Which flavor of ice cream is best?
- What type of stick-on bandage is most comfortable?
- Which kind of laundry detergent cleans stains the best?

TABLE 1.

	Brand A	Brand B	Brand C	Brand D	Brand E
Thickness					
Price					
Absorbency					
Tearability					
Durability					
Amount of lint					

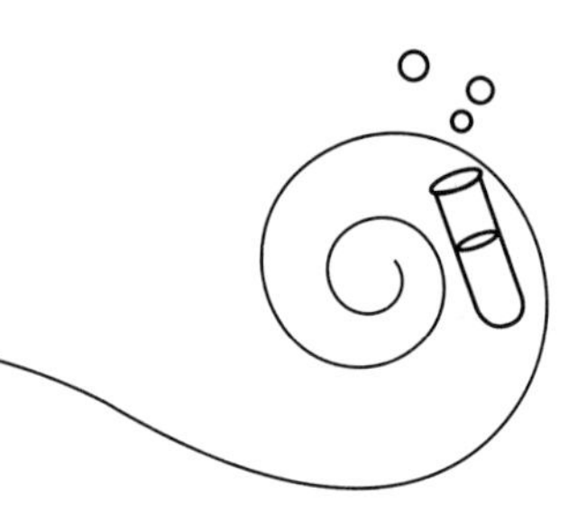

- What brand of dish soap cuts grease the best?

Science fairs should be scientific investigations. Making science fairs more scientific and meaningful is possible in any elementary or middle school classroom. By posing simple everyday questions, investigating them, and answering them, children will become scientists, see how science is applicable to everyday life, and enjoy the process of completing a science fair project. Later, you can introduce more detailed steps, which might include researching the topic and writing a report.

Potpourri

What Have Researchers Been Saying about Science Fairs?

Lawrence J. Bellipanni and James Edward Lilly

Teachers are always looking for exciting ways to stimulate students' interest in science. One of the most popular methods of doing this has been encouraging participation in science fairs. The science fair has existed for many years and has taken different forms around the world. A common goal for science teachers has been the development of a scientifically literate society as detailed by the American Association for the Advancement of Science (1997).

Science fairs relate to the *National Science Education Standards,* Science as Inquiry Standards. The inquiry standards state that students should "ask questions, plan and conduct investigations, use appropriate tools and techniques to gather data, think critically and logically about relationships between evidence and explanations, construct and analyze alternative explanations, and communicate scientific arguments" (National Research Council, 1996).

Science fairs can and should be run at all grade levels. The earlier students gain hands-on experience with developing simple scientific concepts, the easier it will be for them to later perform more complex studies in science.

If students follow the scientific method as they carry out experiments for their science fair projects, it will help them understand scientific concepts, and with proper guidance from their teachers, it will lead them to a lasting interest in both science and engineering.

Most teachers are exposed to the science fair, either in school or in a methods class in their senior year of studies. Their experience, regardless of how they become aware of it, will dictate whether or not they will carry out such activities in their future classrooms.

To help stimulate more interest in science fairs, we offer a brief history and rationale of the science fair, discuss the research that has been done on science fairs, and describe how science fairs have grown to become a truly international event.

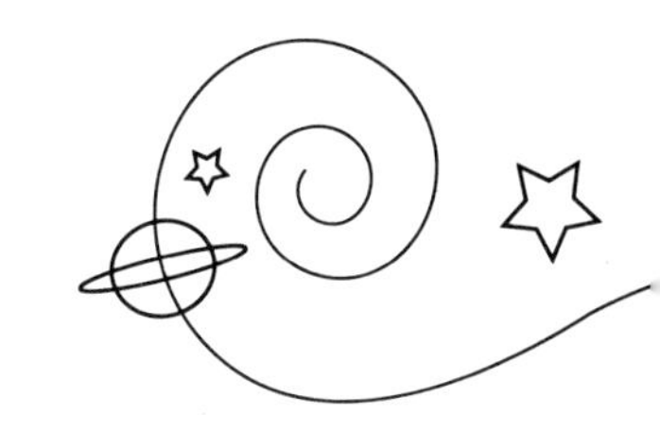

Historical Background

Science exhibitions originated in 1828 in New York with the Science and Technology exposition held by the American Institute of Science and Technology. This institute had a tradition of providing public expositions of science and technological innovations such as Morse's telegraph and Bell's telephone (Silverman, 1986). An important aspect of these exhibitions was the awarding of medals to those inventors and scientists who had demonstrated distinguished work at the exhibit.

In 1928, the institute began to shift its focus away from industrial expositions toward the stages of a children's science fair held in cooperation with the American Museum of Natural History. The 1928 fair is regarded as the first student science fair, and it is the model for all subsequent science fairs (Silverman, 1986). Eventually, the industrial fair evolved into the International Science and Engineering Fair (ISEF) that is held today.

Established in 1921, the Science Service of Washington, D.C., formed science clubs throughout the United States. It was a nonprofit organization, and many industrial sponsors contributed to its success. The major objective of the Science Service was to promote and popularize science. The Science Service, aided by G. Edward Ferdrey of Westinghouse, organized and sponsored the first Westinghouse Science Talent Search in 1942. According to Otis Allen (1954), former superintendent of Lefiore County Schools, the first National Science Fair was held in Philadelphia in 1950.

Throughout the 1950s, science clubs and science fairs continued to grow. With the success of the national science fair, plans were made for an International Science Fair. The first ISEF was held in 1964 in Seattle (Bellipanni and

The competition factor of science fairs is usually reduced at the elementary level to help students feel self-confident and encourage science learning.

Brown, 1986-1987). Finalists from 208 affiliated fairs across the United States and abroad participated, representing the work of over one million students. This fair, as well as the fairs that followed, was dedicated to inspiring greater interest among students in the fields of pure and applied science.

Rationale for Science Fairs

A common rationale for science fairs cited throughout the literature is that science fairs provide the student with an opportunity for hands-on research and learning. Students demonstrate the ability to identify a problem, formulate a hypothesis, determine a procedure, gather data, interpret the results, and draw conclusions (McNay, 1985).

The competition factor associated with science fairs is usually reduced at the elementary level to help students feel self-confident and encourage the learning of science. Typically, all involved in the elementary science fair receive some type of reward to reinforce the idea that

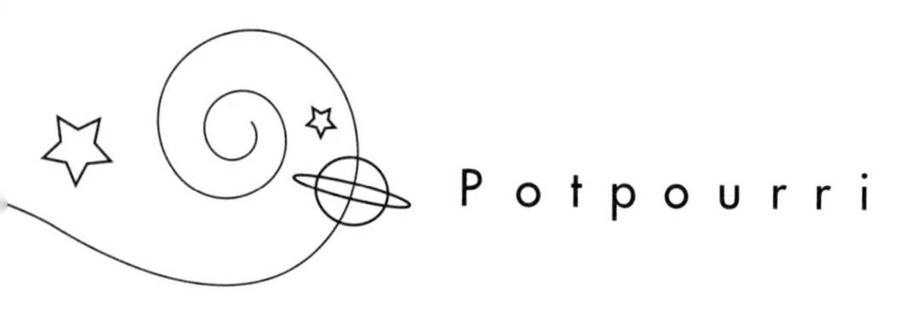

students who do science fair projects are winners. A certificate of participation, medal, or some other reward encourages students to continue to participate in science fairs. "The mere formation of a problem is far more often essential than its solution, which may be merely a matter of mathematical or experimental skill. To raise new questions, new possibilities, to regard old possibilities, and to regard old problems from a new angle requires creative imagination and marks new advances in science" (Albert Einstein as quoted by Wolfe, 1987).

International Popularity

Science fairs are growing in popularity in the United States and around the world. Participation in the ISEF has reached an all-time high.

At the 44th ISEF in 1993 held in Biloxi, Mississippi, 831 participants representing 416 affiliated fairs exhibited (Science Service, 1993). The 1993 fair included projects from 46 states, the District of Columbia, American Samoa, Guam, and Puerto Rico. Sixteen foreign nations were also represented: Argentina, Bolivia, Brazil, Canada, Chile, Denmark, Finland, France, Germany, Ireland, Japan, New Zealand, Republic of China (Taiwan), Sweden, United Kingdom, and Uruguay (Science Service, 1993). Nine of the 16 foreign nations were added during the 1992–1993 school year along with American Samoa, giving the fair more international balance.

Judging Science Fair Projects

Lagueux and Amols (1986) point out the difficulties in being a fair and impartial science fair judge. These two science educators developed a computer program to aid in science fair judging. This program is no longer available, but a more recent computer program, developed by the Mississippi Science and Engineering Fair, is available for use in evaluating and scoring science fair projects. The program interprets scores awarded to each project by the science fair judge and determines a relative scoring level for each judge. (The program costs $250 and is available from W.J. Sumrall, MSEF, Box 9705, MS State, MS 39762.)

Edelman (1988) suggests that, where possible, judges should be recruited from a local college or university science faculty, science-related business organizations in the community, or local science organizations. Judges should be chosen by a committee made up of the previously mentioned groups of scientific personnel. Judges should be given instructions prior to reviewing projects and interviewing students; if possible, judges should be given scoring sheets along with information on the criteria for scoring a project. It is further emphasized that judging may be either an individual or team effort. The basic criteria outlined by Edelman for judging projects are

- creative ability,
- scientific thought and/or engineering goals,
- thoroughness of work, and
- skill and clarity of project display.

Time limits should be established by science fair coordinators to avoid one student getting more attention from judges while other students are shorted.

Mandatory Science Fair?

In the literature, no author has recommended mandatory student participation in science fairs. The National Science Teachers Association (NSTA) position statement on science competitions (available at *www.nsta.org/position*), approved in 1986, supports only voluntary student participation. In the position statement, NSTA recognizes that many kinds of learning experiences, including science competitions, can contribute significantly to

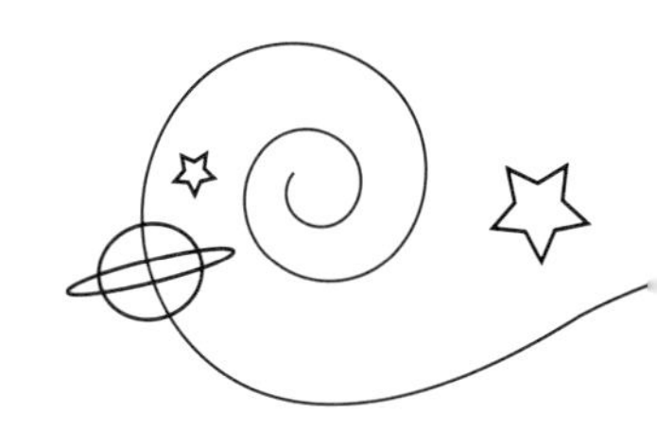

the education of students of science.

With respect to science competitions, NSTA takes the position that participation should be guided by the following principles:

- Student and staff participation in science competition should be voluntary
- Emphasis should be placed on the learning experience rather than on the competition
- Science competitions should supplement and enhance other educational experiences
- Emphasis should be on scientific process, content, and/or application
- Projects and presentations must be the work of the student with proper credit to others for their contributions

Tohm (1985) states that science fairs are a major campaign to increase student skills and to allow students a chance to have fun with science. Long-range planning is necessary to make the science fair experience worthwhile. Though a successful science fair requires an enormous amount of time and energy, the payoffs are impressive—students get excited, parents become involved, and school-community relations are improved as the community is invited to take part in making the fair a success (Hansen, 1983).

Hansen attributes the success of the science fair to planning done by a committee of parents and teachers. The scheduling of monthly meetings and the effective recruiting of more volunteers to work the program lead to bigger and better science fairs.

The library can help young scientists in preparing for a science fair and offers the students a wealth of inspiration and information. Libraries can help by

- updating the library's collection of science titles,
- establishing special circulation rules preceding the science fair,
- setting up schedules,
- offering an inservice library session for teachers,
- introducing or reviewing library and study skills,
- recording projects in progress,
- correlating science and language arts,
- opening science centers, and
- displaying winning science projects and awards.

To conduct and maintain a science fair, there are certain procedures that teachers must go over with their students. Teachers must go over each step that the student should follow in completing a science fair project. It is important to understand the established guidelines for the sci-

When students learn to adequately set up projects and communicate the project findings accurately, the quality of their projects decidedly improves, and this allows students to think like scientists. The elementary teacher has to be the guiding light for the student.

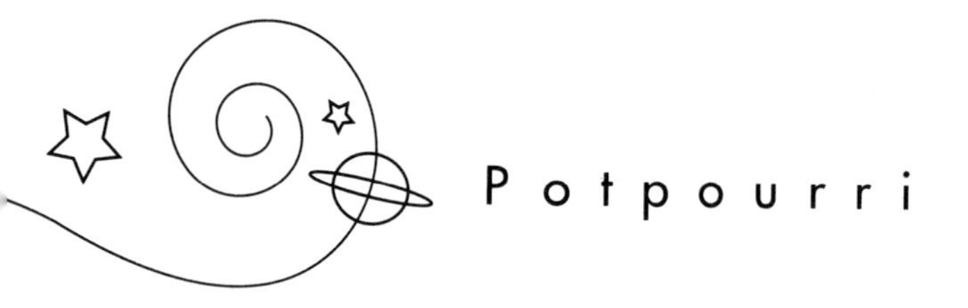

ence fair and review these with the students so that they follow the rules throughout the science fair competition.

Texley (1984) points out that most students have little or no experience in the art of "putting together the science fair project." Science fair projects should begin with the curiosity of the child and in turn foster wonder and amazement in the individual. Texley summarizes the qualities that make the best projects:

- stretch students' investigative skills,
- enable students to question the mysteries of the world, and
- delight students in understanding the complexities of life.

Corner (1984) points out that students must be careful in judging the effectiveness of their research projects. They often arrive at misconceptions about their project results. Corner emphasizes that knowing the right way to evaluate and interpret results can allow students to come up with really sparkling independent research projects.

Teachworth (1987) stresses the importance of comprehensive project goals and a stringent schedule discipline. When students learn to adequately set up projects and communicate the project findings accurately, the quality of their projects decidedly improves, and this allows students to think like scientists. The elementary teacher has to be the guiding light for the student. Teachers help students with library research and also make available all resource materials the student may need to be able to develop a credible idea for a project to be entered in the science fair.

A Rewarding Experience

Science fairs should be a rewarding learning experience for all, especially for the children who participate. Students should be encouraged to do their best and should be given credit for trying. Not everyone connects with science right away. A lot of adults still have difficulty with science, but perhaps this could have been averted if their exposure to science had been an early one. Some students take time to learn anything, and naturally this requires a lot of patience on the part of the teacher.

With an early introduction to science, the student may have a good chance of doing well in science later on in life. Science fairs can be instrumental in making science make sense to children who may otherwise miss the opportunity to learn more about the world in which they live during those precious elementary years.

After being involved in science fairs for over 35 years, I feel that I have seen many different types of projects. However, the science fair projects that stand out in my mind are those done by elementary students where the students actually become part of their project by dressing to fit the part. For example, one year several students created a project on honeybees. One student dressed as the queen bee, another as a worker bee, and the last as a drone. This level of personal involvement and excitement about science is one of the science fair's greatest rewards.

References

Allen, O., Roberts, R.C., and Sheely, C.Q. (multiple interviews 1953–1954).

American Association for the Advancement of Science. (1997). *Update 1997: Project 2061—Science Literacy Changing for a Future.* Washington, DC: Author.

Bellipanni, L.J., and Brown, F.W. (1986–1987). Abstracts of Mississippi Science and Engineering Fair Finalists and Alternates (Vols. 83–89).

Corner, T.R. (1984). Do mouthwashes really kill bacteria? *The Science Teacher, 51*(6), 34–42.

Edelman, P. (1988). Science fairs: The ways and hows. *Updating School Board Policies, 19*(2), 1–3.

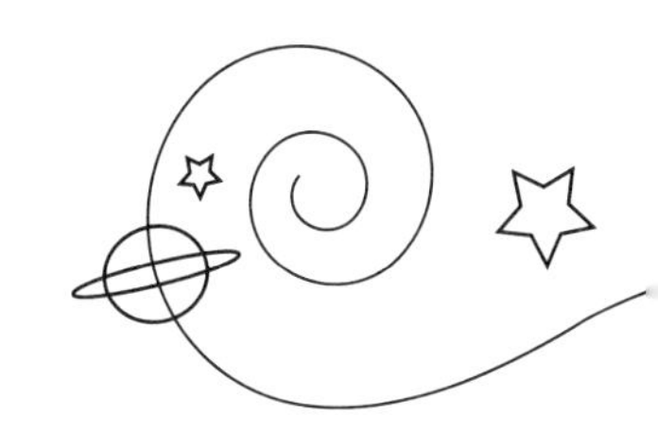

Hansen, B.E. (1983). Planning a fair with a flair. *Science and Children, 20*(4), 9–11.

Lagueux, B.J., and Amols, H.I. (1986). Make your science fair fairer. *The Science Teacher, 53*(2), 24–28.

McNay, M. (1985). The need to explore: Nonexperimental science fair projects. *Science and Children, 23*(2), 17–19.

National Research Council. (1996). *National Science Education Standards.* Washington, DC: National Academy Press.

Science Service. (1993). *Abstracts, 44th International Science and Engineering Fair.* Washington, DC: Author.

Silverman, M.B. (1986). Effects of science fair project involvement on attitude of New York City junior high school students. *Dissertation Abstracts International, 47,* 142A.

Teachworth, M.D. (1987). Surviving a science project. *The Science Teacher, 54*(1), 34–36.

Texley, J. (1984). How to create problems. *The Science Teacher, 51(1),* 28–31.

Tohm, B. (1985). Mission possible: The science fair. *The Science Teacher, 52*(9), 40–41.

Wolfe, C. (1987). Search: A research guide for science fairs and independent study. Zephyr Press. (ERIC Document Reproduction Service No. ED 317 373)

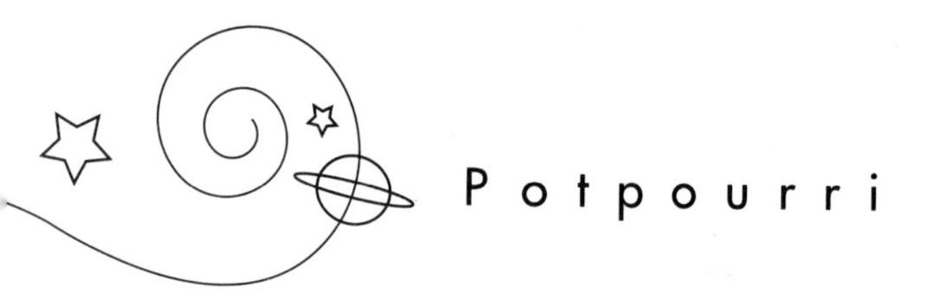

Whoever Invented the Science Fair...

Linda H. Sittig

Whoever invented the science fair was certainly not the parent of a first grader. I speak with the voice of experience, having recently survived our family's first encounter with a science fair.

When our six-year-old daughters announced that they wanted to enter a project in their school's science fair, I inwardly cringed. Remembering how difficult math and science had always been for me, I had a feeling that this project could turn out to be a colossal headache. The enthusiasm on their faces, however, made me ignore the little voice inside me begging to say no, and I asked them what project they had in mind.

I would probably have favored some of the experiments with magnets I remembered from my elementary science days or perhaps a project on different types of clouds. But the girls said they wanted to build their own bird feeders and paint them different colors to see if birds liked one color more than another, and I asked myself how complicated that could possibly be. We would construct two bird feeders, spray paint them, add birdseed, and then watch to see where the birds would eat.

However, since my husband is a junior high science teacher, he had his own ideas about how a science fair project should be carried out. Sitting down as a family, we began to discuss the battle plan, and I heard him uttering phrases like "data retrieval charts" and "correct observational procedure." Worse, he went on to say that our daughters would need to weigh and record the precise amount of seeds, in grams, that the birds consumed each day. I raised some objections based on the academic capabilities of six year olds, but I finally had to capitulate and agree that if you decide to enter a science fair, you do the project right, or not at all—even if you are only six years old.

Needless to say, all free time during the next two weeks was consumed by the project, and I for one felt enormous relief when the finished product was deposited on a table at the science fair. But looking back now, I realize that my children benefited from this experience in a number of tangible ways. They practiced math skills by weighing the birdseed daily on a balance scale (and exercised their fine motor coordination as they adjusted the weights of the balance). Large muscle control came into play as the girls banged away with hammers, and lan-

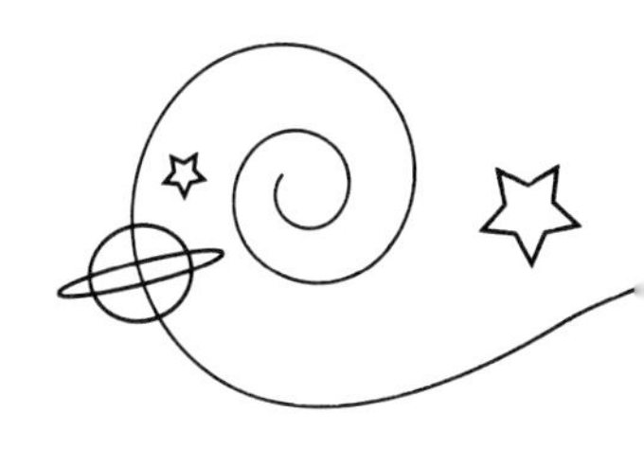

guage skills increased as they read books on how to build bird feeders and how to feed birds. They even practiced their handwriting because they kept a daily log (AKA data retrieval chart) of their activities. Most important, they had the experience of getting an idea and following it through to its conclusion.

I learned a valuable lesson too: children are often capable of achieving more than we give them credit for. Although our daughters could not have handled their science experiment without adult supervision, the credit for conceiving and carrying out the project belonged to them. And this realization suggested a response to the dilemma that troubles many parents and teachers. How can we challenge children without applying too much academic pressure?

The key to the success of the science fair project seemed to lie in the fact that our daughters took the initiative—it was not forced upon them—and they were given the freedom to pursue their own idea. Of course, children do not always come up with ideas of their own that they are eager to pursue. But if adults could be more sensitive to what interests children or students, we could guide—not push—them into activities that they would find challenging and exciting. And if we could be more sensitive, we would nurture, not discourage, the inquisitiveness without which children will never reach their full potential.

Science fairs do come only once a year (I'm still thankful for small mercies), but there are many other opportunities for us to encourage our children's minds to grow. We have only to take advantage of them.

By the way, red seems to be a very popular color with birds.

Science Fairs? Why? Who?

Evelyn Streng

"What's good for high school science is good for elementary science!" Is it? The attention given to science fairs at the junior and senior high school level has led to interest in and emphasis on holding fairs at the elementary school level. This trend has led elementary educators to consider the value of their use.

Opinions differ—as indicated by varying practices of "to have or not to have." Those who have not reached a conclusion would do well to recognize that the elementary science fair should (a) consider the nature and development of the elementary school child, and (b) involve projects that serve the highest objectives of science education.

Child Development

Studies suggest that the elementary child is curious, and that natural curiosity can be directed to scientific investigation. Joseph H. Kraus, a noted science fair coordinator, says: "Beginning science interests peak at age 12, with age 10 now coming a close second. Better than 10 percent of the nationally recognized students are launched toward a scientific future before they even enter kindergarten."

Although some childhood interests flower early, it is important to recognize the differences in developmental patterns. Also, perhaps only a few pupils in a grade-school class may be science oriented. Some creative, talented children may not have the patience or persistence demanded by "scientific investigation."

A fair amount of guidance and direction for the child-investigator is necessary, for the extent to which an elementary pupil can independently develop a project is questionable. The following criteria are desirable when deciding if participation of children in a science fair is appropriate.

1. Only children with a genuine interest in a science project and the initiative to see it to completion without undue adult prodding should be expected to participate in a science fair. A science project should never be a requirement for a class or a necessity for a good grade in science.
2. Any judging of a science fair project or display should consist of helpful comments and suggestions rather than comparative ratings or prizes. If projects are shown in one place, the emphasis should be on the stimulus of shared interests rather than on competition between classes or schools.

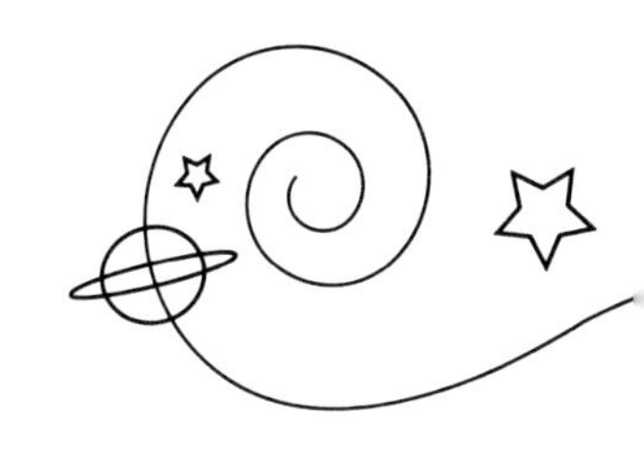

Suitable Science Projects

Suitable science projects are those which increase and direct a child's interest and competency in science. Worthwhile projects are those which are *problem-centered* and in which the *process* is important—not those that center on showmanship or gadgetry. Some categories of appropriate problems for an elementary science fair are

1. ***Observation of the environment***
   - What kinds of trees seem to grow best in our area?
   - What living things may be found in a cubic foot of garden soil?
   - How do some insects change as they "grow up"?

   These are the simplest types of problems, involving a study of the surroundings to classify and organize what is there.

2. ***Demonstration of a basic principle of science***
   - How does electricity travel?
   - What causes erosion?
   - How does a machine make work easier?

   These are not really "research problems," for the answer is known at the start. Their value is in enabling the student to clearly explain a basic idea.

3. ***Collecting and analyzing data***
   - What is the average October weather like in our town?
   - What is the rate at which a pet drinks water?
   - How does the number of seeds produced by different plants compare?
   - Is there a relationship between the phases of the moon and the weather?

   In this type of problem there is no manipulation of nature by the student, but there are directed and recorded quantitative observations. This is more specific than simple observation, which is merely descriptive. Computation of averages, ratios, and rates and performance of other analytic processes will be part of this type of project.

4. ***Controlled experimentation***
   - What is the effect of temperature on the activity of mealworms?
   - What is the effect of the moon phase on the germination of seeds?
   - What difference does the kind of wire make in the resistance of an electric circuit?

   This is the most valuable type of problem from the viewpoint of understanding science. It involves the use of controls—situations identical except for the one variable under consideration. Quantitative aspects are surely implied. It is apparent that the "answers" to some problems (e.g., "What difference does the kind of wire make?") are known to scientists, but they will be unknown as far as the children are concerned.

It is quite possible that elementary children may come up with some original problems to which answers will not be found in the science text. In the execution of a project, children may make the valuable discovery that they do not have sufficient evidence for a valid conclusion. A science project which concludes, "This experiment does not show any relationship between A and B; more experiments are needed" may be as meaningful as one which comes to a remarkably demonstrable "answer."

Shall we have an elementary science fair? Only if careful consideration is given to the nature and the needs of students and to the objectives to be accomplished!

Science Fairs for All:
A Science Fair Project with a Diverse Group of Young Learners

Donna Gail Shaw, Cheryl Cook, and Teralyn Ribelin

Imagine overhearing the following from a kindergarten, first-, or second-grade student attending a school science fair: "That wasn't a science fair project because they already knew the answer before they did the experiment." "Teacher, look at my brother's project. [pause] Where is the hypothesis? He would have gotten a blue ribbon if he had had a hypothesis." "They wrote down their observations, but nobody can read them!"

These comments were made by students from a K, 1, 2 multi-age, full-inclusion classroom after they had spent a month conducting their own scientific investigation. In previous years the teachers had taken the K–2 students to the school science fair to look at all the projects; but, the experience was not meaningful because the students and teachers knew little about how scientific investigations were conducted. However, this year after doing their own experimentation, the students knew the purpose and procedure for a scientific investigation and recognized when other students followed proper methodology for good scientific inquiry. The teachers were impressed with the students' level of thinking and intense curiosity; the experience was enriching for students as well as teachers.

Topic: student equity
Go to: *www.scilinks.org*
Code: SFP40

Beginning Steps

The students involved in this project attended an urban school in the Anchorage School District. The 52 students in the K, 1, 2 multi-age inclusive classroom (21 kindergarten students, 20 first-grade students, and 11 second-grade students) spent the entire school day together. Seventeen students needed special education services, seven students used English as a Second Language services, six children functioned above grade level, and one was gifted.

Two master teachers used a coteaching arrangement the entire school day. The support staff included one morning kindergarten aide, two special-education teacher assistants, a half-day special-education teacher, and a speech and language specialist. For the duration of the science fair project, two elementary science practicum students from a local university and their science methodology professor assisted with science planning and instruction (e.g., attendance at all weekly planning sessions, weekly supervision of students in the science learning center, responsi-

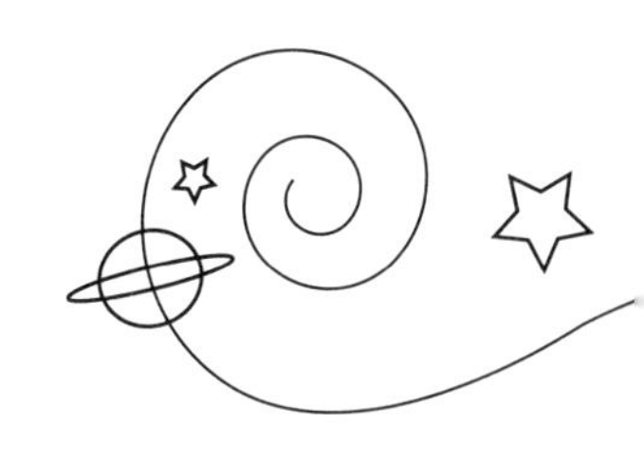

FIGURE 1. Scientific method.

Certain features characterize good scientific inquiry. The following components, commonly known as the scientific method, help students apply these features as they conduct scientific investigations.

Question: State what you want to investigate in the form of a question.

Research: Find out as much about the topic as you can before conducting the experiment.

Hypothesis: Predict what you think will happen in the experiment.

Procedure: Provide a sequential explanation needed to do the experiment. Establish the variables important to the investigation. The experimental variable is the one condition that you change in the experiment. It is the factor that you are comparing or testing. The controlled variables are the conditions that need to remain the same during the experiment so that they do not affect the results.

Results: Record the data obtained from the experiment. Journals, charts, and graphs are typically used to display results.

Conclusion: State what the investigation showed. You will accept or reject your hypothesis. It may also include other explanations, such as conditions that you were not able to control that may have affected the results.

bility for teaching science lessons one day a week).

The philosophy that *all children can learn* served as the basis of the present project: Children learn in different ways and at different rates, and they need to be actively engaged in their own learning. All children are valued as unique individuals with strengths and talents to share with others, and all children are respected as contributing members of a caring classroom community.

A Science, Technology, and Children unit—Plant Growth and Development (National Science Resource Center, 1992)—was used as part of the required science curriculum for the school district. Students learned about the complete life cycle of a plant including germination, growth, and development and learned the importance of recording observations in writing and with scientific drawings. Throughout the unit a science learning center provided students many opportunities for observing plants and recording observations.

The university professor taught weekly hands-on lessons to the students to help them understand the scientific content related to their science fair project (e.g., plant parts and functions, transportation of liquids, photosynthesis, conditions necessary for healthy plant growth). The curriculum addressed the *National Science Education Standards'* K–4 Life Science Content Standard C: All students should develop understanding of the characteristics of organisms, the life cycles of organisms, and organisms and environments (National Research Council, 1996).

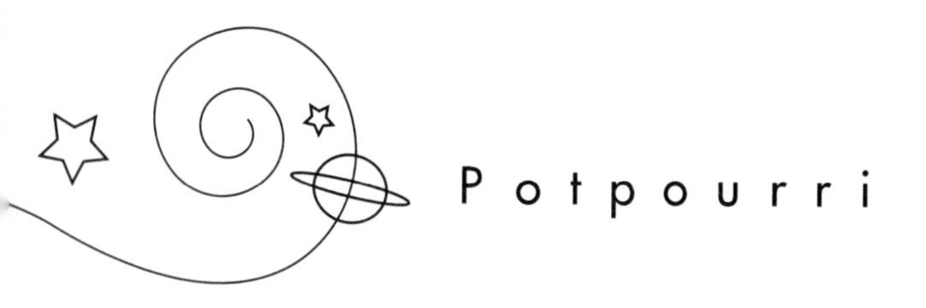

The Project

At a team planning session, the classroom teachers asked the science education professor what she thought about teaching the class how to do a science fair project. The ensuing discussion marked the beginning of an exciting, meaningful learning experience for both teachers and students. During the project students learned the scientific method of problem solving (Figure 1 on previous page), used this method to conduct a scientific investigation pertaining to the plant growth and development curriculum, and entered the results of the scientific investigation in the local and state science fairs. The project addressed the K–4 Science as Inquiry Content Standard A: All students should develop abilities necessary to do scientific inquiry and an understanding about scientific inquiry (National Research Council, 1996).

Asking a question. During the first lesson teachers explained to students that scientists must have a question about something before they can do an experiment. Through their plant unit activities, students had learned the importance of water in the healthy development of plants. To help students find their own question, the teachers led a class discussion that focused on a comparison of liquids that humans drink (soft drinks, milk, orange juice, coffee) and the water that plants take in. During the discussion, the students brainstormed a long list of liquids that humans drink. The teachers wrote all the names of the liquids on the chalkboard. The students had been taught that plants need water, and as soon as the teachers switched the discussion to plants, the students began to wonder "Can plants take in liquids other than water and still grow?" This question led to the following investigation.

The teachers divided the class into six cooperative learning groups, considering the following factors: ability level (mixture of high-, medium-, and low-performing students), personality, language development, gender/ethnic balance, and reading ability (at least one strong reader in each group). Each team chose a team name such as Biologists, Experimenters, and Safe Workers.

Formulating a hypothesis. In the next lesson students learned that a scientist usually has a suspected answer to the question he or she has posed or a prediction of what will happen: a hypothesis. This hypothesis is based on the scientist's past experience and what he or she knows about the topic. Since it was not feasible to test all the liquids brainstormed by the students, names of liquids were written on pieces of paper and six of these liquids (one for each team) were randomly selected (see materials list, below).

After each team chose a liquid, the teachers read the ingredient labels found on the products to the students and the students discussed what they knew about the liquid. (The day before the lesson, the teachers had gathered the liquid products the students had brainstormed.) Each student in the team made a hypothesis about how the selected liquid would affect the plant's growth, and then the team members voted on a hypothesis for the team as a whole. (The students' hypotheses stated that a certain liquid would either cause the plant to 1) grow better than the plant that takes in water, 2) grow the same as the plant that takes in water, or 3) not grow as well as the plant that takes in water.)

Each team had a folder with the team's name printed on it. Individual and team hypotheses sheets were placed in the folders. All papers were kept in these folders until journal notebooks were decorated and collated at the end of the project.

Gathering materials. Before beginning the experiment, the teachers gathered the following

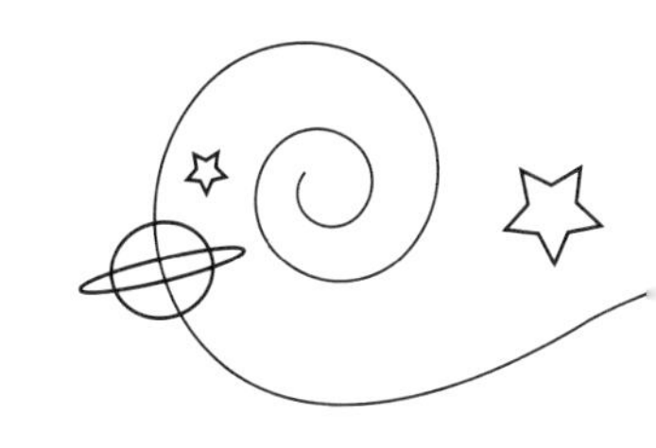

materials (If you are doing this activity with older students, they can help collect these materials):

- 12 plants (Plants should be the same kind and approximately the same size; tall, straight plants are best. Variegated peperomia plants were used in this project.)
- six plant trays (one tray for two plants)
- six types of liquids selected by students (cola, milk, soft-drink mix, orange juice, antacid analgesic tablets dissolved in water, and coffee)
- tap water
- clear plastic 8 oz containers for liquids
- camera and film
- paper for recording data
- materials for science fair display

(The Parent-Teacher Association paid for the plants and the university assisted with other materials, such as trays and cups.)

Procedure—Designing the experiment. In this lesson the term *procedure* was introduced to the students as meaning what scientists do to get information to answer the question—some scientists refer to it as "testing the hypothesis." The class discussed the importance of keeping the procedure the same every time and the importance of consistency among all teams. This discussion included identifying the experimental variable (type of liquid) and the control variables (e.g., same amount of liquid each time, same amount of light, constant temperature, same type of plant). The procedure was established and displayed in the classroom as follows:

- Give one plant 60 mL of water.
- Give the second plant 60 mL of liquid selected by team.
- Give plants water and other liquids on Monday and Thursday only.
- Use the data collection sheets (Figure 2) to make any observations every Monday and Thursday.

FIGURE 2. Data collection sheet.

Team Name ______________________

Plant #1 ______________________

(liquid)

Today's Date ______________________

Experiment Day # ______________________

Observe and draw

What we did

Who watered the plant? ______________________

Who recorded? ______________________

Who made the observations?

Changes we observed

____

Conducting the experiment and recording data. Twice a week for one month the students gathered with their team members and one adult to water the plants and record their observations. The adult ensured that all students had an opportunity to participate in the experiment and data collection (e.g., measuring stems, measuring and applying liquid to plant, describing leaves). In addition, photographs were taken several times throughout the investigation.

Interpreting the data and sharing the results. On the last observation day the student teams, with an adult helper, reviewed all previous observations and determined the results of the ex-

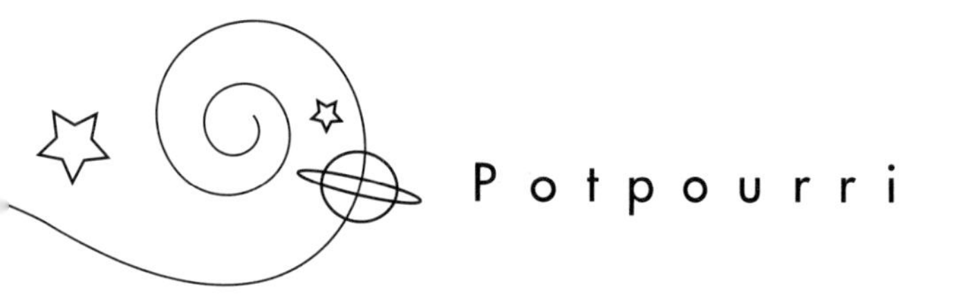

Students wondered what would happen if they put the plants in different locations or if they used more or less of each liquid.

periment. Each team prepared a presentation for the rest of the class. The presentation included what liquid was used, what some of their observations were, what happened (the results), and why they thought it happened. The following include some explanations from students:

- The plant didn't have the right thing to drink. Water is the right thing for a plant to have.
- Kool-Aid went in the leaves and the leaves died. There is lots and lots of sugar in Kool-Aid. Water makes it grow to be strong and healthy. The plant with water is growing because water is good for it like protein is for us.
- The milk spoiled and made the plant die.
- If you want a plant to grow up, give it water.
- Orange juice isn't good for plants, it has something in it that water doesn't. The orange juice got rotten. It got inside the plant and the plant got rotten inside the stems.
- Orange juice went down to the roots, then went into the stems and then into the leaves and the leaves wilted.
- Orange juice is heavier than water so the leaves bent down.
- Coffee doesn't seem to hurt the plant.

Making a conclusion. The final part of the experiment involved making a conclusion. In this lesson the students learned that a conclusion was a statement made by the scientist about what the experiment was able to show. The students looked at their original hypotheses, determined if they were correct or incorrect, and discussed why the results supported or did not support their hypotheses. They discussed their observations of the plants and compared them. Were the plants different? Why? What did we do differently with each plant? Students completed this aspect of the project with an adult.

After each individual team member made a conclusion, they voted on a team conclusion, which was then shared with the class. By majority vote, the students decided that the class conclusion should be that plants could not take in other liquids and still grow. An exciting student-initiated discussion ensued with students wondering "What would happen if... ?" By the end of the discussion, the teachers had recorded over 40 new questions that students wanted to answer.

The students wondered what would happen if they tried other liquids than the six that were selected. Many of the children stated the names of the other liquids they wanted to try. They also tended to focus on the controlled variables. For example, the students wondered what would happen if they put the plants in different locations (in the refrigerator for the plant with milk so it wouldn't spoil; in different areas of the classroom closer to the window or farther away) or if they used more or less of each liquid.

Designing and making the science fair display. After completing the scientific investigation, students were placed into teams based on the skills needed for each task associated with preparing the science fair display. Teams

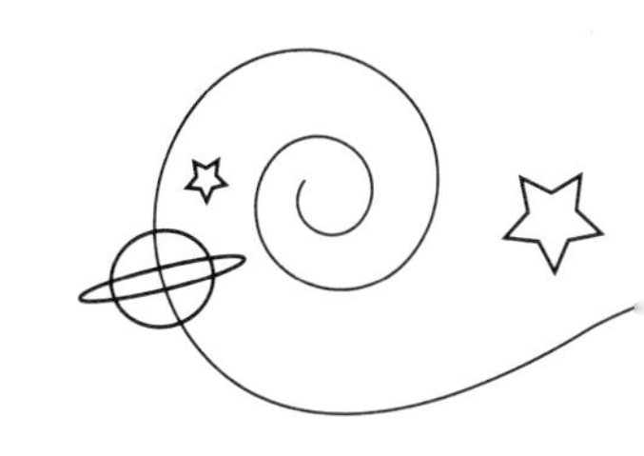

Example of a Completed Science Report

Draw a picture of a plant.
Label the parts of the plant.

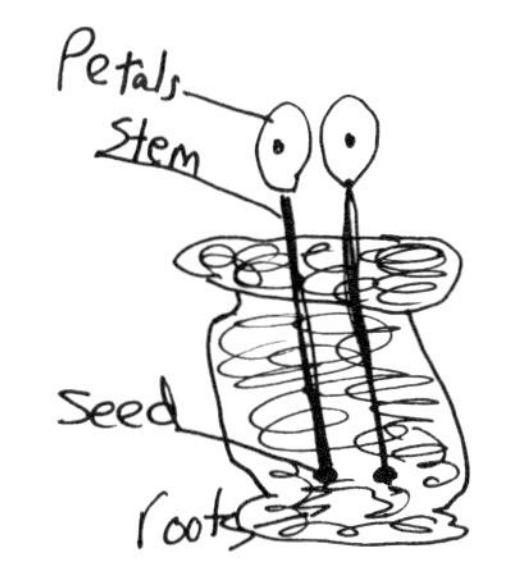

Write about our science project.
HINT: Don't forget the hypothesis.

My hypothesis was that the alkasezer Wouldgrow better than the Water,

What are 3 things you learned in science?

I learned about a hypothesis and a Conclusion. I learned that Plants cannot drink other liquids.

Write about how a plant grows.
HINT: Start with a seed.

first youplantthe Seed Then you water it. Then you give it some sun and then you water.

Why is a flower important?
HINT: Who visits a flower?

The rain and the Sun and you do.

worked with adults to complete such tasks as deciding which information should be typed and included on the backdrop display, decorating the display board, designing journal covers and organizing the data into sections, and composing the science fair project abstract.

Assessment for science fair project. Assessment was an ongoing task throughout the entire plant unit. Numerous adults observed students as they participated in the hands-on lessons, role-plays, and discussion groups. The teachers looked for the development of process skills: observing, inferring, communicating, hypothesizing, interpreting data, controlling variables, and investigating (use of scientific method).

After completing the project, first- and second-grade students produced a science report that included such information as a written description of the science project, three facts they learned in science, how a plant grows, why a flower is important, and a labeled drawing of a flowering plant (see "Example of a Completed Science Report," above).

All students were individually interviewed regarding their understanding of the scientific method. Students were given four pictures representing the question, hypothesis, experiment, and conclusion. They placed these pictures in a logical order and verbally explained what they knew about the scientific method. One explanation included a student saying "You get an idea, you decide what you think is going to happen, you do the experiment, and then you tell a lot of adults what happened and what you learned."

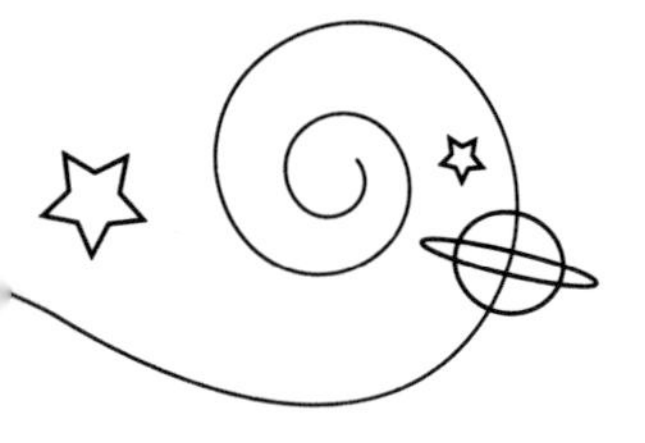

FIGURE 3. Recommended scoring rubric.

______	up to 2 points	student placed the pictures in a logical order
______	up to 4 points	student verbalized the correct process
______	up to 3 points	student accurately used all the scientific terms
______	up to 1 point	student added additional information
______	Total	

Score interpretation

8–10 excellent 5–7 satisfactory below 5 needs improvement

This was a practical way to assess each student's level of understanding (see Figure 3).

The science fair. The specific guidelines for the state science fair were followed and students were invited to attend the fair to participate in oral interviews with science fair judges. Initially the students were nervous about the interviews; however, it wasn't long before the students were describing their project to any adult who would listen. The project won first-place ribbons at the state science fair and the local school science fair.

Final thoughts. The teachers spent time together reflecting on what they thought contributed to the success of this science fair project with such a diverse group of students. Their thoughts are summarized below.

- Students were part of a learning community that valued diversity and individual contribution as a natural part of life. Students supported each other's learning, which reaffirmed a characteristic of multi-age classrooms.
- All students had an opportunity to be successful. Appropriate modifications were made to

TABLE 1. Examples of activities and application to multiple intelligences.

Logical
Sequencing life cycle of plant
Counting leaves
Measuring height of plant and amount of liquid for watering plants
Using the scientific method to answer a question

Verbal
Listening to and reading books about plants
Oral presentation of results of experiment
Using descriptive language while making observations
Oral interviews with teachers and science fair judges

Visual
Designing and making display board and covers for group folders
Recording observations through drawings

Bodily
Participating in role-plays (life cycle of plants, how plants move liquids through tissue, etc.)

Interpersonal
Working in small groups to complete a scientific investigation
Discussing project with group members weekly
Using Kagan Cooperative Learning Strategies (1997), such as Numbered Heads Together

Intrapersonal
Reflecting on individual roles and their importance to the success of the science fair project
Celebrating the achievement of a blue ribbon and complimenting each other and teachers for the role each played in making the project a success

Naturalist
Planting seeds
Identifying plants
Learning interdependence of animals and plants

Musical
Exploring water cycle through a round

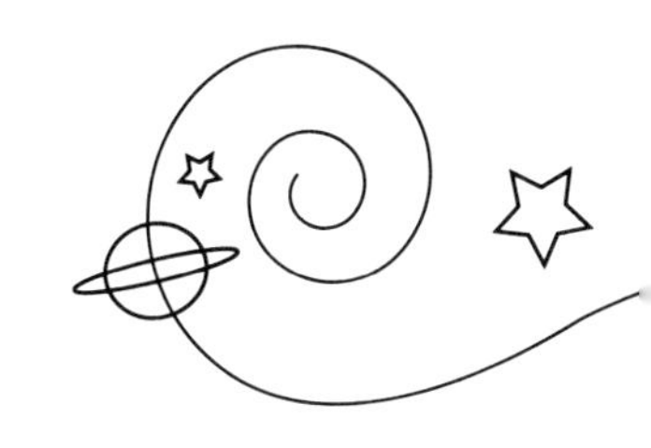

ensure maximum participation and learning by all students.

- There was collaboration among professionals (classroom teachers, special educators, practicum students, university professor), a widely supported component for effective inclusive schooling (Stainback and Stainback, 1990; Thousand and Villa, 1990).
- The adults and students were equally curious and excited about the outcome of the scientific investigation.
- The activity used a variety of instructional techniques: direct instruction, guided discovery, cooperative learning, learning centers, large- and small-group instruction, hands-on activities, extensive opportunities to think and talk about what was being learned, review and reinforcement on a regular basis, role-plays, exposure to nonfiction reading material, and performance assessment.
- The teachers applied Howard Gardner's Theory of Multiple Intelligences (Campbell, 1996; Roth, 1998) to the students' learning experiences. Table 1 provides examples of activities for each intelligence.

And the Plant Grows On

The students learned in different ways and at different rates, but they all learned how to use the scientific method to conduct an experiment. The university professor and practicum students visited the classroom eight months later and were greeted with the following remarks: "My plant is still growing. I only give it water." "I told my grandma not to put milk on the plants in her garden!" "I'm still a scientist!" "Can we do another science fair project this year?"

Resources

Campbell, L., Campbell, B., and Dickinson, D. (1996). *Teaching and Learning Through Multiple Intelligences.* Needham Heights, MA: Allyn and Bacon.

Kagan, S. (1997). *Cooperative Learning.* San Clemente, CA: Kagan Cooperative Learning.

National Research Council. (1996). *National Science Education Standards.* Washington, DC: National Academy Press.

National Science Resource Center. (1992). *Plant Growth and Development: Science and Technology for Children.* Washington, DC: National Academy of Sciences.

Roth, K. (1998). *The Naturalist Intelligence.* Arlington Heights, IL: SkyLight Training and Publishing.

Stainback, S., and Stainback, W. (1990). The support facilitator at work. In W. Stainback and S. Stainback (Eds.), *Support Networks for Inclusive Schooling: Interdependent Integrated Education* (pp. 37–48). Baltimore: Brookes.

Thousand, J.S., and Villa, R.A. (1990). Sharing expertise and responsibilities through teaching teams. In W. Stainback and S. Stainback (Eds.), *Support Networks for Inclusive Schooling* (pp. 151–166). Baltimore: Brookes.

In the Balance

Lawrence J. Bellipanni, Donald R. Cotten, and Jan Marion Kirkwood

You have just hung up the telephone after a brief conversation with the science teacher at a local junior high school, and somewhere along the line you've "volunteered" to be a judge for the school's science fair. Suddenly you are responsible for evaluating projects that students may have spent months working on and for deciding which projects are best. Making these decisions is no easy task, but if you keep a few points in mind, you can turn your judging duties into a rewarding experience for both you and the students.

Regardless of the grade level you're working with, you should note the quality of the work the students have done and determine how well they understand their projects. The project should include research, experimentation, and application—not simply library work. But as you apply these standards, always consider the grade level of the student whose project you're judging and the general level of expectation for that particular fair.

Here are some specific criteria to use:

1. Creative ability. Has the student shown intelligence and imagination both in asking the question and arriving at the answer? Is the student original in deriving and applying data? Did he or she build or invent any equipment to use in the project?

Remember, anyone can spend some money, but it takes a creative person to devise the equipment needed for a particular project. Ask students where they got their ideas. Creative students are always coming up with new twists to old ideas; such ingenuity indicates that you're dealing with an interested young scientist. Collections may show diligence, but they seldom show creativity. So don't be tempted into giving them high marks unless they have some true scientific merit.

2. Scientific thought. Is the problem stated clearly and unambiguously? Did the student think through the problem and pursue his or her original question without wandering? Was the experimental procedure well defined and did the student follow each step toward the expected outcome?

Did the student arrive at the data experimentally (as opposed to copying them out of a book)? Are the data relevant to the stated problem? Is the solution offered workable?

3. Thoroughness. A solid conclusion is based on many experiments, not a single one. Does the project test the main idea of the hypothesis? How complete are the data? How well did the student think through each step of the experiment? How much time did he or she spend on the project? There are few loopholes in a project that has been done thoroughly. Ask the student questions about the project to determine how well he or she understands the problem.

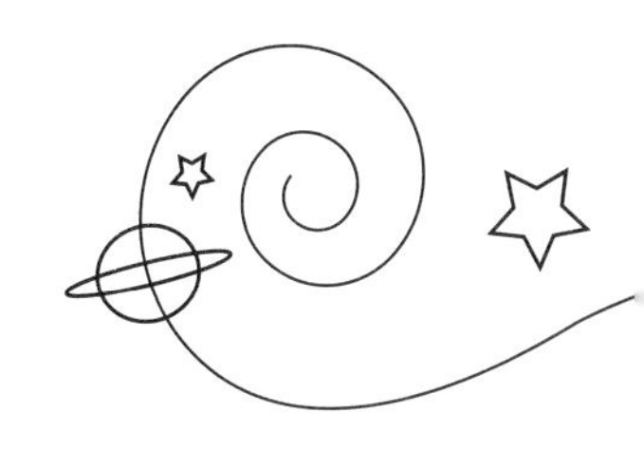

Suddenly you are responsible for evaluating projects that students have spent months working on and for deciding which ones are best.

4. Skill. Since you don't know the students personally, you will need to have some way of determining how likely it is that they did the work themselves. Ask them if they had any help with their projects. (But use common sense here. If the project requires using an electric saw and the student is in third grade, it would be permissible—indeed advisable—for an adult to perform this task.) You can usually tell how much of the actual work students have done by observing them while they demonstrate or explain the project.

5. Clarity. The project should be set up so that the judge can follow the procedure and understand the data without getting confused. Students should have written the data clearly, using their own words, and they should be able to discuss any portion of the project. The main purpose of the project is to show that students can formulate, test, and present research.

Though these five criteria are basic, the standards for judging particular science fairs may vary, depending on the grade level of the participants or the types of projects involved. The teacher supervising the science fair should make certain that each judge has a judging sheet, indicating not only the criteria to be used but the points that each item is worth. If you do not understand one of the criteria, ask the teacher or coordinating judge for clarification before judging begins. Your responsibility to the children is to be as fair and objective as possible, and that can happen only if all the judges use the same criteria in the same way. And remember: each child's project is very important to that student. So whether the project merits a blue ribbon or not, be sure to provide proper encouragement so that students will continue to investigate their own ideas.

Science Fair Fatigue

Cecelia Cope

Several years ago, following our annual science fair, I noticed my students were burned out. When it was all over, most of my students felt more drained than challenged or exhilarated. Even students who received top awards at the state science fair doubted they would compete next year.

To boost classroom morale after the science fair, I decided to create an activity with interdisciplinary tie-ins. My students needed to express their science fair experiences, both positive and negative, and in the process achieve a sense of closure. Here's how the activity works.

I provide three possible titles to get students started: "Science Fair: The Experience"; "Science Fair: The Good, the Scary, the Fun"; and "Science Fair: It's Not Just Science." Students can either use these titles or create their own. Once students have selected a title, they spend the rest of the class period composing poems, writing essays, or creating artwork. I encourage students to spend as much time at the supply table as they need. Here they have access to drawing paper, construction paper, scissors, glue, magazines, markers, colored pencils, poetry guidelines, and sample projects from previous years. But first I ask students to think about the entire science fair process, from start to finish. I ask them to recall how they researched and selected their topics; learned how to use science fair indexes and other reference tools; and visited zoos, museums, or universities to gather information.

As they think back, I ask them how they felt when they had difficulty choosing a topic or when they experienced setbacks while putting together their projects. Some good probing questions are "What was it like being surrounded by books and not knowing where to begin?" "Did you want to give up when you had difficulty scheduling a phone interview?" and "How did you decide on how to revise your experiment after something didn't go as planned?" I also ask students to think about the judging process: "What was it like to read the judges' comments and answer their questions?"

I give students several minutes to think about their individual experiences and choose what stands out in their minds. If what stands out is mostly negative, I encourage students to identify something positive to balance out their overall experience.

Although most are able to recall both positive and negative aspects of the science fair experience, all students benefit from taking part in a

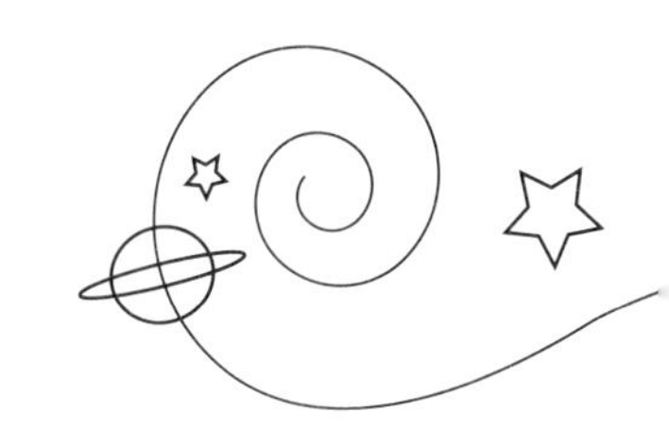

group discussion to identify particular feelings and reactions that they may have forgotten. For example, my colleagues and I hosted several help sessions to assist individual students during the planning stages of the science fair. By discussing different aspects of the science fair, students were better able to assess the overall experience.

The range of students' work is always remarkable. In the past, students have made elaborate collages and drawn detailed cartoons. Students have also incorporated several styles of poetry into their projects. Best of all, the assignment gives students an opportunity to share their feelings in a productive way, without the fear of receiving a low grade. And giving students permission to express their feelings positively affects their attitude toward science and renews their interest in learning.

Even though students are asked to voice both positive and negative experiences, just expressing themselves profoundly affects students' attitude toward science. Student participation in regional and state competitions has increased, and more and more students approach investigative activities willingly. Because students are encouraged to remember their science fair experiences, they realize that learning is more important than winning.

Points of View

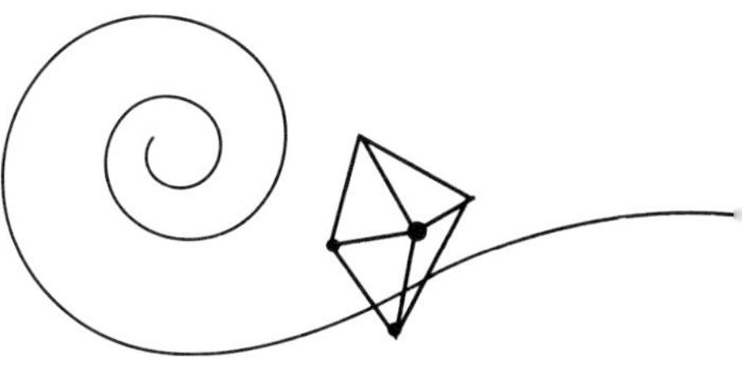

The Need to Explore: Nonexperimental Science Fair Projects

Margaret McNay

I once asked a biologist—an international authority in marine invertebrate embryology—about his childhood experiences with science. In particular, I wanted to know whether he had ever competed in a science fair. Yes, he told me, at age 14 he had entered his school's science fair. Fascinated by the living things he could see through a microscope in a drop of pond water, he had taught himself how to take pictures of them and had then made a display of his photomicrographs.

He did not win.

Some of us are not surprised: science fair projects are supposed to be experimental, to demonstrate that the young scientist can formulate and test a hypothesis, gather data, interpret results, and draw conclusions. Fair projects that display information or demonstrate a principle or process have often been considered insufficiently scientific and have even been described as not only missing the essence of science but also being inconsistent with the goals of teaching science. Such narrow-mindedness arises from a popular but inadequate view of the nature of science. Indeed, nonexperimental projects can evoke the spirit and nature of science as fully as investigative ones.

Topic: scientific inquiry
Go to: *www.scilinks.org*
Code: SFP54

Some fairs are breaking with tradition by accepting entries from both categories, according nonexperimental project separate status and judging them against separate criteria. For the sake of grade school entrants in particular, science fair organizers should embrace this practice enthusiastically. They should not ask the following question (with its implied negative answer) about a display that offers no hypotheses. ...

"Interesting, But Is It Science?"

Instead, organizers could often answer in the affirmative, if they could agree that method alone does not define science. As useful as the scientific method and experimental design are, they do not define the limits of science. Popular opinion and the emphasis of many curriculum guides notwithstanding, science is much more than observing, hypothesizing, measuring, predicting, testing, and concluding. Essentially, science means questioning the world, wondering how it works, and, while delighting in its mysteries, raising hope about the possibility of coming to understand some of them. "What science has to teach us," wrote Jacob Bronowski, "is not its techniques but its spirit: the irresistible need to explore." Tech-

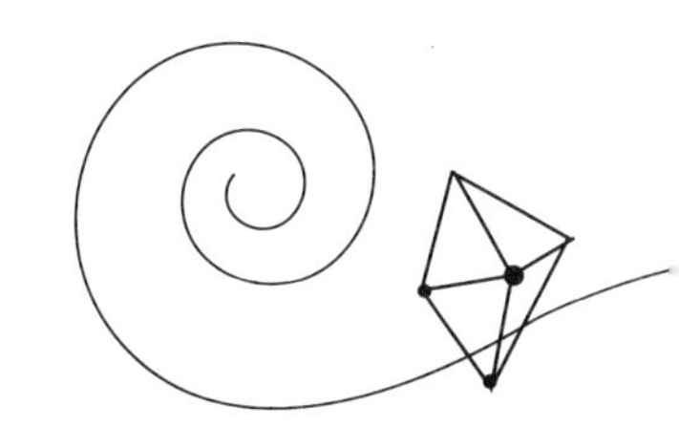

niques are valuable only insofar as they make it possible for us to pursue something that has already aroused our curiosity. When we press children to choose a hypothesis to test or a question to resolve experimentally, unless we are careful, we may emphasize problem solving, measuring, controlling, and predicting at the expense of curiosity and intellectual involvement. Going too quickly and directly to the choosing and testing of hypotheses can bypass the wonder and delight in which science begins and can even deny children the essential experience of science.

Wondering and Questioning

Nonexperimental science fair projects offer children an opportunity to get involved with topics that sincerely interest them, even if those topics have already received a great deal of sophisticated exploration. Free from the necessity of limiting their investigation to a single aspect that can be tested experimentally, children can explore broadly and deeply, to the satisfaction of their own curiosity.

When will they learn experimental techniques? When a question arises calling for such techniques, there will be time enough to master them. And unless wondering and questioning are encouraged as fundamental scientific processes, children won't need experimental techniques anyway: they will develop no deep interests and no abiding curiosity about the world, and they will, therefore, have no questions to ask.

Nonexperimental Approaches

Some of the following approaches can help create an interest in science as profound as that of the marine biologist whose commitment survived losing the science fair prize 30 years earlier. Your students may choose to stick with nonexperimental projects or they may decide to change to an investigative mode. They may also want to combine their approaches. You can help them to learn that science offers an exciting universe enormous enough to accommodate inventors, observers, philosophers, describers, experimenters, and more. Help their horizons widen through encouraging them to work on projects such as the following (partially adapted from Evelyn Streng; see page 38 of this book):

1. Presenting three-dimensional displays, reports, and posters based on literature searches. Many topics in which children are interested do not lend themselves to direct observation but can teach through being represented in objects, models, and diagrams. For example, students can make models of
   - the universe
   - the structure of atoms and molecules
   - the ocean floor
2. Building working models or presenting technical demonstrations. Children can come to understand many scientific principles through learning about technology. For example, they can
   - make a home-built seismograph
   - show how a rocket works
   - explain how nylon is made
3. Demonstrating a basic scientific principle. Even if children already know the answers to a question when they start to work, explaining the principle involved and showing how processes occur deepen their understanding. For example, students can show
   - what causes wind and water currents
   - what causes erosion
   - how machines make work easier
4. Observing the environment. Studying questions like the following can involve the chil-

dren in a study of their surroundings that leads them to begin to classify and organize what is there. For instance, they can learn

- what lives in a drop of pond water
- how some insects change as they grow
- what kinds of webs different spiders make
- how plants disperse their seeds

5. Collecting and analyzing data. In these problems, children do not manipulate variables but go beyond the descriptive to make and record observations and analyze data. They can find
   - how fast bean plants grow
   - how the number of seeds produced by different plants compare
   - how efficient various pulley systems are

The Joy of Learning

Nonexperimental projects do tend to dominate elementary school science fairs, but this fact is nothing to deplore. Though experimental designs with properly controlled variables and appropriately drawn conclusions are important, they are not inherently more valuable than the sense of wonder and enthusiasm for finding out about the world. Children need to learn experimental techniques, yes, but they also need to explore.

The marine biologist who didn't get a ribbon at 14 continues to do essentially nonexperimental science. While his current work is more sophisticated than his childhood photographs of pond water creatures, it is essentially similar in nature. His descriptive studies of the invertebrate larvae that live in water help biologists better understand the complex mysteries of cell differentiation and the development of organisms. Much of his work does not require experimental methods, yet it is accepted as significant research. Why shouldn't we also recognize nonexperimental projects at the science fair?

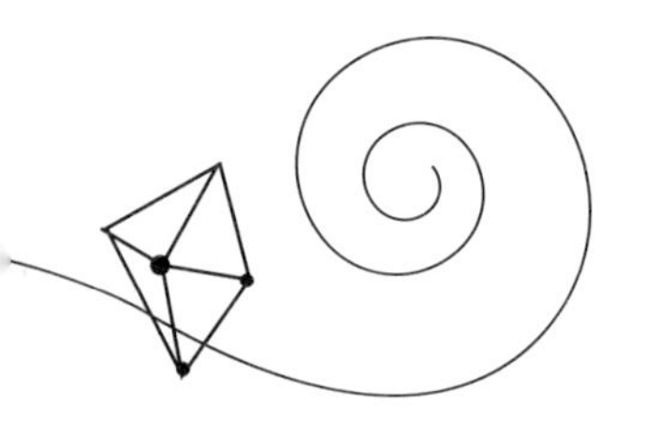

Why Science Fairs Don't Exhibit the Goals of Science Teaching

Norman F. Smith

One need attend only a few elementary- and intermediate-level science fairs to discover that they are all more or less alike. A second discovery will follow close behind: most of the projects in these fairs have little relevance to the goals of science teaching. From my long experience as a scientist, plus many assignments as a science-fair advisor and judge, I suggest that the cause of this situation lies where no one may have thought to look—in the way science fairs are operated and judged.

An analysis of the kinds of science fair projects we see time after time at the elementary and intermediate levels clearly shows a disparity between the goals of science fairs and those of science teaching. Nearly all fair projects can be placed in one of the following five categories:

1. Model building (for example, the solar system, volcanoes, clay models of frog organs)
2. Hobby or pet show-and-tell (for example, arrowheads, slot cars, dogs, baby chicks)
3. Laboratory demonstrations right out of the textbook or laboratory manual (for example, distillation, electrolysis, seed germination)
4. Report-and-poster projects from literature research (for example, fossils, birds, bees, the astronauts, the ear)
5. Investigative projects that involve the student in critical thinking and science processes, such as measuring, reducing data, and drawing conclusions (for example, tests of reaction time, effectiveness of various detergents, comparison of the performance of vacuum bottles with insulated jugs)

If the goal of science teaching is to improve skills in model building, library research, poster making, or following laboratory-manual directions, then projects from the first four categories are appropriate. Such projects may in-

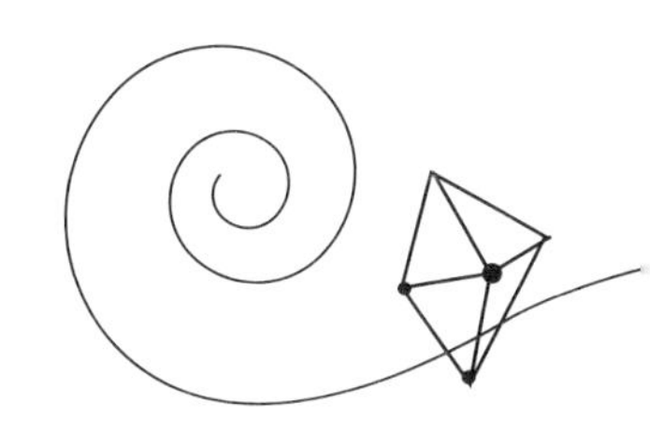

deed stimulate students' interest in science and increase their knowledge of science, in addition to contributing to social and communication skills. But if one of the primary goals in science teaching is to teach critical thinking, inquiry, and investigative skills, then projects in the first four categories simple do not match this goal, or are, at best, ineffective approaches to it. The essence of science is found only in category 5, in which the student must conceive and plan a project, perform an investigation, and analyze data to arrive at some conclusion or some new understanding.

Problems Worth Investigating

This being the case, why is it that projects in the first four categories (models, posters, show-and-tell, and laboratory demonstrations) are predominant at science fairs, while projects dealing with discovery and investigation are decidedly scarce? This is a question long overdue for investigation; it applies as well to extra-credit projects and normal laboratory activities.

Discussing this question with science teachers yields a number of viewpoints. Some teachers blame the poor science backgrounds of those in their own profession, especially among elementary teachers. Because elementary teachers may be neither highly skilled in science, nor entirely certain about the goals of science teaching, they tend to be more comfortable with activities closely allied with bookwork. Students, too, are more comfortable with projects that can be lifted from books than with less familiar and more original projects that probe the unknown.

Other teachers point out that the kinds of projects currently popular represent a "point of entry" into science for the younger student. This viewpoint has validity, and some use of these kinds of projects is undoubtedly justified.

Why is it that models, posters, show-and-tell, and laboratory demonstrations are so predominant at science fairs, while projects dealing with discovery and investigation are decidedly scarce?

As it turns out, however, most students remain stuck with these projects year after year, repeating selections from the first four categories until they move into high school. Then the rules suddenly change, and only original experimental or technical projects generally are accepted, at least at major competitions. What are missing or at least underemphasized in the present system are *transition* projects, in which the student moves from the easy poster project to a deeper look at the science aspects of his topic, and finally to sampling the process of investigation. (Indeed, some of the exotic projects in high-school science fairs are more oriented to technology than to science; consequently, one wonders whether these students have ever had the experience of designing a simple experimental project.)

Many teachers are aware that the question of what kinds of projects might or should be done

in science fairs is almost never discussed. In particular, there is little or no discussion or agreement beforehand among teachers, students, and science fair judges as to the *purposes* of the endeavor and the *criteria* by which entries will be judged. As a result, teachers find themselves coaching students in the execution of projects that will be judged by persons unknown to them, and according to criteria that are not carefully considered by those making the rules.[1]

In my view, the most startling reason for the present emphasis on noninvestigative projects is the orientation of the judges themselves, which causes them to reinforce projects in the first four categories and to discourage investigative projects. While this orientation may seem contrary to the interests and instincts of scientists, it is really quite understandable, given the usual conditions of scientists' involvement in science fairs.

View from the Scientist

The lot of the scientist asked to judge a science fair is not a happy one. Armed with a specialty in some branch of science, but often with little or no knowledge of science education, he surveys, clipboard in hand, a scene that is quite foreign to his professional world—a vast arena of eager students of widely varying competencies, who are presiding over projects that vary even more widely in quality and science content. The casual observer may marvel at the diversity of projects he sees, but the judge has the grim job of sorting out these projects and finding some basis for declaring a few of them the "winners."

How "winners" are often chosen is best illustrated by true examples from a junior-high science fair. On one table the judge finds a project labeled "Air Pressure," with a Bunsen burner and three or four gallon cans that have been crushed by air pressure. There are also two other gadgets right out of the lab manual neatly hung on ringstands so that people can blow into them to demonstrate "air pressure" for themselves. The posters are adequate; the students are responsive to the questions of the judge and impress him with their understanding of the topic.

A few tables away is a project in which two students have compared the ability of a thermos bottle, a plastic jug, and two or three other kinds of containers to keep liquids hot. They have described on posters their purpose and test procedures. Their equipment is on display, and other posters show tables of data and graphs of the variations of temperature with time that the students stayed up half the night to measure. The graphs are neatly made and the data, though a bit rough, look good. Under questioning, the students show that they have drawn some conclusions, but their understanding of the science principles behind what they have done seems somewhat shaky to the judge. He finds that they don't know a great deal about heat, how heat is transmitted, or about insulation.

As the reader by now has guessed, the first-in-class was awarded, for at least the thousandth time in the history of science fairs, to the students who crushed the cans with air pressure. Second and third awards went to excellent library-research-and-poster projects on fish and

---

[1]Judging criteria usually consist of items like: originality, thoroughness, accuracy, clarity, organization, neatness, creativity and skill, dramatic effect, technical skill and workmanship, social implications, communicative skills, science content, and scientific approach. Such criteria are usually vaguely defined for the judges, if at all, and are weighted very differently from one science fair to another. It is amazing that items such as "science content" and "scientific approach" are sometimes omitted or are weighted as little as 10 percent.

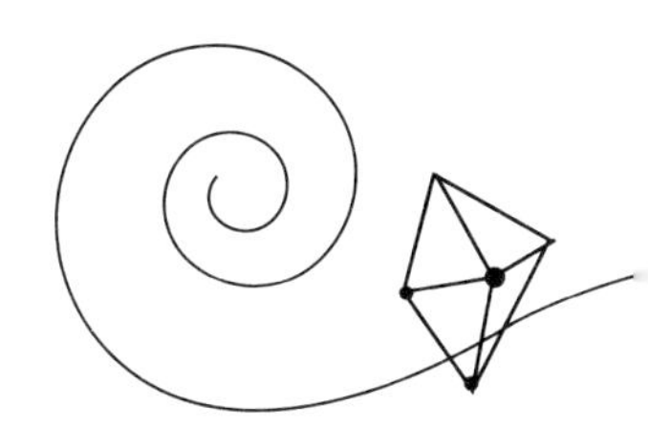

birds, respectively. The investigation of thermos bottles and insulated jugs did not place or receive any recognition.

The process by which the judge had arrived at his decision later became clear from discussions with him. With only general criteria to guide him and a sketchy, at best, understanding of children and science teaching, he had relied on his best instincts as a scientist. The questionable understanding of science principles shown by the thermos investigators troubled him—perhaps "repelled" would be a better word—and kept him from seriously considering their project, whereas he was drawn toward the apparent competence shown in the demonstration-and-poster project. A "good" understanding of some concepts of air pressure and a "good" laboratory *demonstration* were more worthy, according to his standards, than a "fair" ability to conduct an *investigation* backed by only a "fair" understanding of the concepts involved.

But what could be wrong with that? We all know that competence is vital in science, don't we? And don't we also try to teach proper understanding of science concepts and principles?

"Investigation" Rebuffed

The consequences of such decisions by science fair judges, however, are obvious. The students who made earnest—and perhaps fruitful—attempts to explore the unknown with an investigative project have been rebuffed. If they try at all next year, they will probably seek a project along the safer lines of library research or laboratory demonstration in which they may, through book learning and practice, acquire the aura of competence for which the "system" has shown clear preference.

Lest anyone think this an isolated event, I cite two other projects from the same fair. One is a

The first-in-class was awarded, for at least the thousandth time in the history of science fairs, to the students who crushed the cans with air pressure.

demonstration of refraction, well-executed by a pair of confident students. They know the principle of refraction and demonstrate it in half a dozen ways using a slide projector and an aquarium, along with several drawings. Their competence in this topic is impressive

On the next table is a project on chickens and eggs, an interest the student brought from home. With some guidance from the teacher, the student undertook to measure the thickness of eggshells from different kinds of chickens. At first, she couldn't locate a micrometer, but did have a feeler gauge. She used an old automobile spark plug, gapping it by trial and error to fit each shell, then measuring the gap with the feeler gauge. Later she was able to locate a micrometer and used it to check and refine her earlier measurements. Her display was unimpressive, her manner shy, but from her data one could learn about the range of thickness of egg shells, and—a most interesting item—the dimensional tolerance within which the chicken manufactures the shell. (Anybody out there know that?)

She also showed a comparison between the thickness of a standard shell and a "soft-shell"—a defective egg occasionally laid by some chickens. She understood well the dietary deficiency that causes soft-shelled eggs and what to do about it.

But in the dazzle of textbook competence from the next booth, the spark of inquiry glowing among the eggshells was unnoticed by the scientist—indeed, may have been snuffed out by his lack of interest and his final decision. First place went to the demonstration of refraction, second place to the familiar "How Seeds Grow," and third place to the ever-popular "How the Ear Functions." Should the student who struggled with eggshells be brave enough to do another project on chickens—or any other topic—next year, she will put curiosity aside and generate instead the most elaborate library-research, poster, show-and-tell project she can muster. Who could blame her?

The dominance of non-investigative projects in today's science fairs suggests that fairs have drifted far from the avowed goals of science teaching. A fresh examination is needed to bring the goals of science fairs and science teaching back together.

Here's one modest proposal: If fair sponsors were to set up a separate judging category for investigative projects, they would immediately motivate students and teachers to move in this direction by guaranteeing recognition of such projects. Over a period of a few years the present monotonous fairs might begin to evolve into new "discovery fairs" in which students, teachers, and the public would discover the adventure of investigation and experience the true meaning of science.

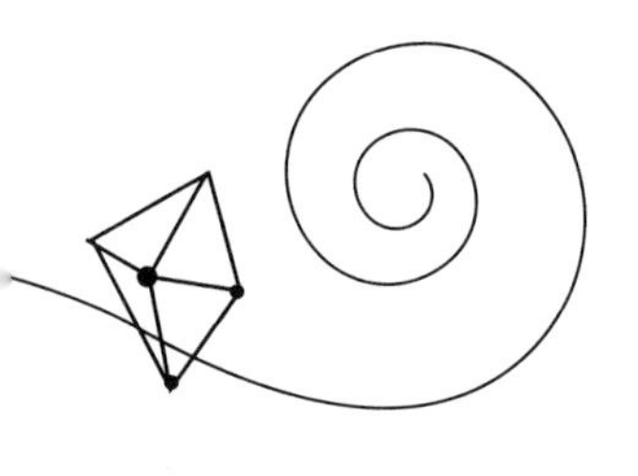

The Trouble with Science Fairs

John Stiles

The sixth graders at a nearby elementary school hold an annual science fair, and, since I hadn't been to one in some time, I was eager to attend while I had the chance. There was an aura of excitement in the building—I could hear the buzz of the crowds in the classrooms as I walked through the halls, and I felt myself lifted by the energy. In the exhibition area, I was impressed by the number of student projects and the scores of adults moving from display to display. The atmosphere was positive and upbeat, full of enthusiasm.

Student scientists were eager to share information and began discussing their projects as soon as visitors approached. My first encounter was with a boy seated behind a table on which rested a metal "Engelspiel" with four candles. The device is a German Christmas ornament that rotates above candle flames. The student told me that he had investigated the relationship between the number of burning candles and the number of revolutions made by the Engelspiel. He was surprised to find that lighting two candles produced less than twice the number of revolutions produced by one candle. Likewise, three and four burning candles did not result in three and four times the revolutions produced by a single candle.

Science fairs promote interest in science, but sometimes they aren't everything they're cracked up to be. What's often missing is the real science of inquiry and investigation.

Although the experiment was simple, it left me with a good feeling: A sixth-grade student wondered about a physical phenomenon, tested it, and came up with a conclusion.

The Downside

As I moved from student to student, however, my initial optimism began to fade. Rather than

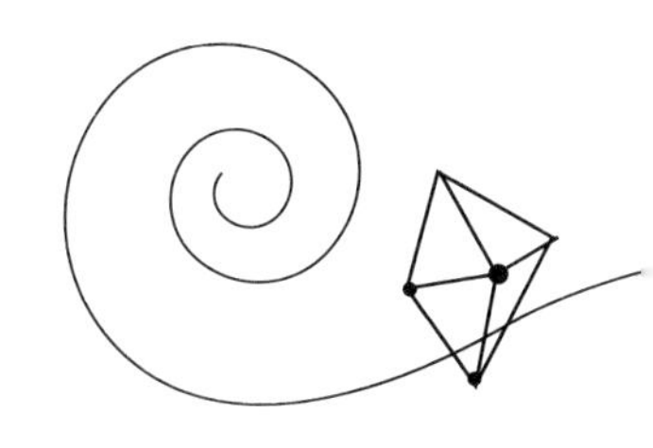

developing and carrying out scientific investigations, the vast majority of participants had simply mimicked activities they found in books. Explanations generally discussed the mechanics of the "experiments," but rarely involved scientific understanding of the phenomena. The misinformation was staggering. Some examples:

- The architect of a model oxygen molecule explained, "When electrons go one way, the neutrons go the other way."
- A student who grew plants in light and dark environments didn't know why one grew better than the other. "Maybe it was water," he offered.
- As he demonstrated a model submarine sinking and rising in an aquarium, a young scientist freely confessed that he had no idea why it worked.
- A floating needle moved in a curved direction to the poles above a submerged magnet. When I asked its creator why it moved in such a path, she began digging in a folder and replied, "I don't know, but I have the information right here."
- The obligatory volcano was quite a production, complete with a smoking crater and glowing interior. When I asked its builder why the two sample rocks had such different characteristics, he informed me, "One is regular and the other is another kind."

A Light in the Dark

I was about to leave when I noticed a boy sitting by himself, surrounded by graphs and charts. Most of the visitors politely passed him by. I talked to him and discovered that he had tested the growth rate of bacteria and had produced a beautiful exponential graph of the results.

It occurred to me that the reason few stopped to talk with him was that his display was not cute or flashy and he was rather shy. But he also had produced a real scientific investigation. Unfortunately, people are often made uncomfortable by this, particularly if they do not understand the concepts involved.

Toward Scientific Science Fairs

It's evident that this "science" fair was far from scientific. On the positive side, it was a popular event, looked upon with great favor and anticipation by students, parents, teachers, and administrators. It certainly promoted excitement about science. On the other hand, it also perpetuated a false idea of what science really is.

As is the case in many schools, these students obviously were excited about their fair and they genuinely liked science class and their teachers. This enthusiasm should be directed toward a working knowledge of the scientific process so that students can discover the joy and satisfaction of investigation and problem-solving.

Many might ask, "Can sixth graders carry out scientific investigations?" Of course they can—and they do. In the same school district, I was a judge at a sixth-grade fair that allowed only scientific investigations. Unfortunately, for a variety of reasons, most students never get this opportunity.

Teachers should keep in mind that investigations needn't be complex to be effective. Students can learn from simple experiments as long as they pursue problems in a scientific manner. For students to begin to understand the work of scientists, they must learn to investigate natural phenomena by a problem-solving process that includes observing, questioning, predicting, testing, collecting data, summarizing, discussing results, and drawing valid conclusions. This can be done at a very basic level, using simple, inexpensive materials. Or, even better, students can

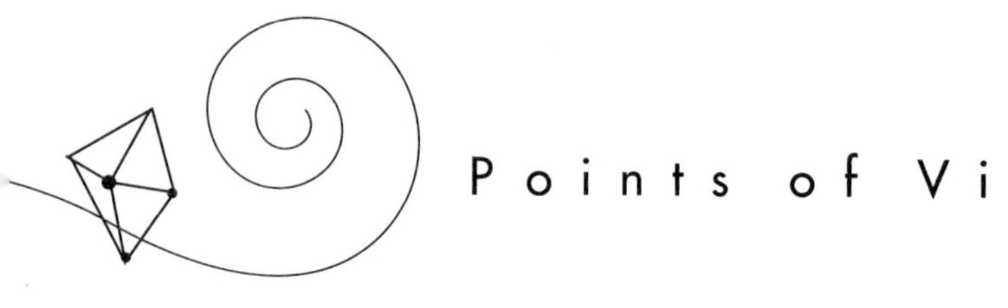

observe and investigate natural phenomena all around them: Simply stake out an area of study in the schoolyard.

Teachers can encourage this inquiry process by allowing students to devise and carry out simple tests for their problems. In addition, we should always ask questions to facilitate inquiry, show interest and enthusiasm, discuss progress, share ideas in group settings, brainstorm, and use as many hands-on activities as possible.

Learning centers allow students to pursue interests independently, and guests from related science fields can provide new information, help foster interest in a subject, and create career awareness. Let the resource materials help students (and you!) learn: No one expects a teacher to be a walking encyclopedia, but he or she should be able to guide student learning.

Science fairs should be scientific. If displays, demonstrations, or models are to be a part of the fair, they can be judged as categories separate from the investigations. In all cases, student research must be closely monitored to ensure understanding.

Making science fun and meaningful by providing a stimulating atmosphere and encouraging students to design simple experiments is possible in any classroom. Help is as close as the teacher down the hall, ideas from journals, local experts in industry and education, and most importantly, your own creative mind. Offer students the opportunity to *be* scientists rather than just read about them and think of the excitement and learning your next science fair could generate!

Section II

Expos and Festivals

Expos and Festivals

Science Fairs: Tired of the Same Old, Same Old?

Debbie Silver

About seven years ago, when I was a teacher at Logansport Rosenwald, a small, rural K–6 school, I was given the job of revamping our science fair program, which at the time had a very limited scope and appealed to few of our students. I wanted very much to find a way to get more students involved and make the science fair a fun community event. Believing that children learn best while having fun and that extending their school experience into the community validates students' hard work, I wanted to involve our small community in the event. I also wanted to foster a positive relationship between the school and local businesses by involving them in the program as well.

There is something for everyone at the Science Expo. Come take a closer look at this science fair alternative.

Competition vs. Cooperation

By reading journal articles and anything else I could find on the subject, I learned of many excellent alternatives to science fairs developed by schools across the country. Seeing that many of these alternatives maximized cooperative learning and minimized competition, I decided to conduct an independent research project exploring the effect of science fair competition on student learning. I found that while competition drives some students to succeed in science fairs, it can be extremely stressful and intimidating for many others, making then believe that science projects are a chore and keeping them from enjoying science. It can also have a negative impact on parents who feel ill-prepared to help their children compete in a contest that often affects the students' science grade for the period.

After all of my research and reading, I was left with so many great ideas for successful programs that I just couldn' t decide among them, so I decided to do them all. I combined what I considered the best of the best, preserved the opportunity for students to enter independent projects for judging, and added cooperative and

noncompetitive options. Thus the Science Expo was born.

The Science Expo is an all-day event meant to show students and the community that science is fascinating and fun. Students can choose to participate in any one of several areas: the traditional science fair, the Share Fair, class demonstrations, the Invention Convention, the Family Physics Fun Festival, and the Family Science Olympiad. The expo also includes business exhibits and special presentations given by members of the community to show how science plays a role in all of our lives. (For detailed descriptions of each area, see "Expo, Expo, Come One, Come All," on following page.) Only students who choose to enter the traditional science fair are judged competitively. The other areas are noncompetitive and in many instances students may work together.

The Science Expo caught on quickly in Logansport, then in the district, and when I moved to Keithville Elementary/Middle School in 1988, a much larger K–8 school, the program became even bigger. Now the Science Expo attracts over 1,400 people in just one Saturday every February.

Making Change

One of the first steps in implementing this new idea was to educate the teachers and the students about all the options the Science Expo offers. While developing the program, the planning committee and I were surprised to discover that many teachers never mentioned science fair projects until it was time to assign them each January. To help students use the scientific method year-round, we provide inservice education to encourage teachers to consistently model and discuss the scientific method.

For the students, each year in late November or early December, we give an hour-long program in the auditorium to acquaint them with the various choices they have for participating in the Science Expo. Even though participation is not required, most of our students choose to be involved in the expo. All students who participate receive extra credit in science class, as well as ribbons and certificates. A generous business owner even donates skating passes for participants.

The Science Expo has been a real success, thanks to a supportive administrator, an industrious committee of teachers, a dedicated PTA, lots of planning, and a positive attitude. I recommend that neophytes start small and add one or two new areas each year. If you and your school are tired of the same old science fair, it is time to analyze your present program and be creative, innovative, and flexible in designing a new one.

Expo, Expo, Come One, Come All

Bright and early on the morning of the Science Expo, students, families, teachers, and community members gather all around the school, setting up exhibits, making last-minute adjustments on projects, and getting excited about the big event. When the expo begins at 11:00 a.m., students and visitors alike move from area to area looking at the projects, asking questions, playing games, and participating in the various activities offered. Each part of the expo is set up in a different area of the school, and everyone attending receives a schedule that lists the locations of all the events.

Traditional Science Fair

In the science fair area, students display individual projects that use the scientific method. At a typical expo you might see a project comparing carbon monoxide emissions from different makes of cars, a student graphing the effects of radiation on seed germination, or a study of which commercial sunscreen offers the best protection.

Projects are judged the day before based on International Science and Engineering Fair (ISEF) rules, and winning students are eligible to compete in the regional and state science fairs. The traditional science fair usually attracts highly motivated students who have confidence in their ability to do scientific research, talk about their findings, and compete with their peers. Teachers closely monitor students' proposals for projects to ensure that they follow ISEF guidelines. In addition, parents of all students who choose this traditional route attend an evening seminar advising them of strict ISEF rules about safety and required paperwork.

Students who compete in the traditional science fair benefit from the process of designing, creating, and defending their unique ideas.They derive satisfaction from having done a project on their own. Most of them enjoy the dialogue with the expert judges (who are carefully chosen based on science expertise and ability to understand and relate to children) and are further challenged during the judging process. Winning students receive trophies, plaques, and prizes donated by community members, local businesses, and local government leaders. For a 30-minute period during the expo, all students stand by their projects to present them and answer questions from visitors.

Share Fair

Whereas the traditional science fair is necessarily very strict, the Share Fair allows great freedom for students who are interested in doing projects but reluctant to enter the traditional science fair. Some students are uncomfortable presenting alone, while others have been discouraged in the past by a judge who told them that their erupting volcano or solar system model was "not really science." But although these types of projects do not model the scientific method, they are still scientific.

The Share Fair allows students to work

Welcome to the Keithville Science Expo

Schedule of Events

8:00–10:30	**Pancake Breakfast**, sponsored by the FHA Club
9:30–10:30	**Registration and Setup**
11:00	**Expo Begins!** • Science Store (elementary library) • Science Food (concession booth) • Traditional Science Fair (main gym) • Share Fair (auxiliary gym) • Invention Convention (band room) • Family Physics Fun Festival (elementary cafeteria) • Metric Olympics (middle school library) • Exhibits (eagle's nest) • Class Demonstrations (walkways) • PTA Raffle Table (eagle's nest) • Louisiana Blood Center (orchestra room)
11:30–12:00	**Rubber Band Shoot-Out** (breezeway in front of elementary gym) **Invention Convention** participants present (band room)
12:00	**Life Air I Helicopter Lands** (near tennis courts)
12:00–12:30	**Egg Toss Contest** (breezeway)
12:15–12:45	**Tae Kwon Do Demonstration** (auditorium)
1:00–1:30	**Riveting Rocket Show** (outside, behind auditorium) **Share Fair** participants present (auxiliary gym) **Aerodynamics Aloft Contest** (breezeway)
1:30–2:00	**Traditional Science Fair** participants present (main gym)
2:00	**Life Air I Departs**
2:00–3:00	**Awards Ceremony** (auditorium)
3:00	**Project Removal and Cleanup**

individually or with classmates, students from other classes, parents, or community members on science experiments, models, or other demonstrations of known scientific phenomena. All students research their topics extensively and display the resulting projects during the expo. As in the traditional science fair area, students present their displays and answer questions during a scheduled 30-minute period.

One student and his father brought beekeeping equipment, honeycombs, and pictures and delighted onlookers with stories about collecting honey and caring for bees. Four eighth-grade girls did a comparative study of home permanents and captivated visitors with their knowledge of chemical reactions and biological facts. A special education student mesmerized young and old with his knowledge of how resistors and capacitors can be used to regulate the flow of electricity.

The Share Fair usually draws the most participants of all the events because it is noncompetitive, the rules are flexible, and students can work cooperatively with peers, family, or others. The Share Fair also gives students the freedom to explore their interests in science and become "mini-experts" on a chosen topic without the fear of being judged. Many students have gained enough confidence from participating in the Share Fair to move on to the competitive science fair the next year.

Class Demonstrations

For the class demonstrations, each class in the lower grades and each department in the upper grades chooses a topic they have studied and presents several of their favorite activities and demonstrations. Teachers, students, and parents work in shifts to allow visitors to learn about the chosen topics and try out the activities.

Class demonstrations have proven to be a very effective way of involving less-motivated students in the Science Expo. Students who are hesitant about working on an individual or small group project seem much more willing to work alongside their teachers, parent volunteers, and peers in showing what their classes have discovered.

Class demonstrations also give individual classes a focus when studying particular topics. One second-grade teacher whose class chose a demonstration about magnets reported that during their study students would say things to each other such as, "Now Shannon, you've got to understand how the magnetic field works or you won't be able to explain about why the iron filings behave that way. Here, let's go over it again."

Many of the demonstrations center around how science affects our lives, such as a Metric Olympics (math department), the physics of sound (band and music department), and aerobic exercise (physical education department). The art department does a different activity each year involving physical science in art such as pottery making, diffusion of colors, and so on. The home economics department bakes bread to show the importance of yeast and sells

the bread to benefit their FHA club. The language arts department decorates the walls with various types of poetry on scientific concepts. The social studies department makes posters showing how science has influenced wars and other major historical events. And the special education department sets up their animal habitats to introduce the rest of the school to their classroom pets.

Invention Convention

The Invention Convention is a relatively new addition to the Science Expo. The Invention Convention encourages students in grades four through eight to research a need and invent something to address it. A prototype of the invention is displayed at the expo and the inventors present their inventions during a designated 30-minute period.

This event has quickly become a favorite among visitors who enjoy the amazing gadgets that students create, such as flip-top plastic milk jugs, sock sorters, tennis shoes with actual springs in the soles, and automatic cow feeders. The Invention Convention encourages students who like to dream and concoct. It particularly appeals to students with a high degree of imagination and creativity who have little patience for the repetitiveness of a traditional science project. They benefit from doing the necessary research on the problem they want to solve as well as making diagrams, prototypes, and models of their inventions. Students also document their resources and write a report of their findings.

Family Physics Fun Festival

This is one of the most fun and popular areas for expo visitors of all ages. Young and old alike flood the elementary cafeteria where the Family Physics Fun Festival is held. Squeals of delight can be heard as a grandmother bats a balloon across a ping-pong net to her kindergartner grandson. Whoops and cheers ring out as a father wins the candle shoot-out contest. Participants huddle closely around an optical illusion exhibit, trying to find the right perspective.

For the Family Physics Fun Festival, middle school students work in groups of four to six to design and construct a game based on a physical science concept. For instance, the balloon ping-pong game illustrates certain aspects of aerodynamics, a plinko game demonstrates probability, and making aluminum foil boats and loading them with washers shows density, water displacement, and buoyancy.

After enjoying the challenge and fun of a game, visitors have the chance to answer three questions about the physical science concept involved in the game. Groups must research their chosen concept, submit a report, and write questions for different age levels—from young children all the way to adults. Correct answers win candy prizes donated by supermarkets.

Before we created the Family Physics Fun Festival, we had difficulty involving the eighth graders, some of whom felt that they were too old or too cool to participate in the same events as much younger students. But the eighth

graders especially enjoy being in charge of this area and it gives them a great sense of pride. All students who participate are very serious about making each game scientific as well as fun. They critique each others' games and questions to make the festival the best it can be. The cooperation is phenomenal, and the benefits are many. Even the shyest students seem to enjoy the attention they get from younger students, who often react to them as though they were famous scientists—taking pictures of them and asking for their autographs. It is so exciting to see how much the students learn about physical science, planning, and research, having a good time all the while.

Family Science Olympiad

In this area, students pair up with a parent or other adult and compete in lighthearted events similar to Science Olympiad activities. Getting cross-age participants involved has proven to be a great way of generating interest from parents and community members. One father helped his teenage son construct a paper airplane that stayed in the air longer than any other team's. He was so proud that he volunteered to come to school and help the fourth-grade class with their airplane projects.

We plan events that take some knowledge of science and are fun for families to try. Sometimes families work together before the expo, such as in the egg drop competition, and then test their solutions at the expo. Various events are scheduled throughout the day. Winners receive small gag gifts or certificates, but the main emphasis is on cooperation, fun, and learning. This area advances science through family involvement. It attracts students of all ages and brings in plenty of parents, even those who usually don't visit school.

Business Exhibits

In order to further demonstrate how science affects daily life, we invite businesses to set up displays relating the importance of science to their work.

Our local chapter of the National Geographic Society sets up a mock dig so that students can actually experience how archaeologists find and label artifacts. Soon after the Science Expo opens, a rescue helicopter from a hospital lands on our tennis courts. It then becomes an active exhibit for visitors to climb into and examine while its crew explains procedures and equipment. Before the expo is over, visitors can see the helicopter lift off and depart. A ham radio operators' club sets up computers that allow students to track satellites and communicate with people in other countries. The blood center sets up a room for adults to donate blood. Students can watch and learn how blood is collected and preserved. Each year, 25 or more groups set up exhibits that bring science to life.

Also in the exhibit area, the PTA runs a science store that sells small science toys such as magnifiers, magnets, kaleidoscopes, and "zoot" tubes. They also run the concession booth and hold a raffle as fundraisers. Combined with donations from exhibitors, these booths raise about $1,400, which sends two of our students to Space Camp.

Special Presentations

We are always happy when members of the community offer to give fun and educational presentations at the expo. So far we have had a rocket show put on by a high school teacher and his students, a "magical" chemistry demonstration by a college professor, a physics show by a local middle school teacher, a Tae Kwon Do demonstration by students from our school, and a hands-on math circus. These types of presentations further emphasize that science is fun and important in our lives.

Go for an Expo!

Daniel Wolfe

Sometimes new teaching ideas pop up in the most unlikely places. At the NSTA convention in Houston a couple of years ago, it occurred on a shuttle bus. I happened to sit down next to Debbie Silver (author of "Science Fairs: Tired of the Same Old, Same Old"; see page 70 of this book) on the bus leaving the convention center. In the course of our conversation, she told me about a session she was giving on the Science Expo she developed. Inspired by her success in creating an alternative to science fairs, I decided to bring the idea to my own school.

As chair of the science and math department of my high school, I approached my fellow science and math teachers with the idea, and they were willing to give it a try. We decided to invite the fourth through eighth grades in our small school district to participate alongside the high school students. Using Debbie's ideas as a guide, we chose the following for the expo: student projects, science games, teacher demonstrations, and business and organizational exhibits.

Student Projects

The main part of our expo was the student projects. The projects were open ended, with the only requirement that they explain or describe a scientific or mathematical principle or concept. Students could work in groups of up to three,

The science expo idea is flexible and adaptable to your particular situation. Here's how one teacher did it.

which helped them to overcome their fears about entering a project. There were no limitations on size, and in fact we had some rather large projects that truly enhanced the expo. For example, two girls from chemistry class made a huge periodic chart listing uses for each element in the appropriate blocks.

Because we have a combined science and math department, we tried to integrate math into the expo in order to involve as many teachers as possible. However, even though we encouraged students to submit math-related projects, none chose to do so.

It was up to the individual teachers to decide how to encourage students to participate. I gave each student who entered a project two points on their next quarter's grade, and I acted as a resource person. Materials were generally the

students' responsibility, but we did supply some chemicals and arranged for the student council to cover the cost of some special chemicals needed for a project on polymers.

All projects were set up and judged on a Friday afternoon. Students could choose not to have their projects judged, but the majority of projects were submitted for judging. We gave first-, second-, and third-place awards in each of three grade-level categories: elementary (grades 4–6), junior high, and senior high. Although we did not have an honorable mention award, we feel it would be good to give more students recognition for their work. We are also considering a separate math award to encourage students to explore math projects. Judging was based on creativity, explanation of a concept or principle, diagrams and written materials, and thoroughness. Each criterion was worth 25 points.

The expo took place that Saturday from 10 a.m. to 2 p.m. If possible, students attended the expo to explain and present their projects for the public. We announced when a particular project was about to be demonstrated and directed interested visitors to the project's area. As a culmination to the day, we were able to take the winners and their projects to a local television station to be interviewed for a news broadcast.

Fun and Games, Expo-style

For the games area of the expo, the elementary, junior high, and high school teachers each chose events that challenged students to build an object and compete with other students. The elementary grades competed in three events: barge building, structure building, and can car race. For the barge-building event, students tried to build an aluminum foil barge that would hold the most pennies without sinking. For the structure-building event, students used plastic straws and pins to see who could build the tallest structure. For the can car race, students built cars from a #10 can with a rubber band propulsion system and competed to see which car could go the farthest.

The junior high and high school students showed interest in only one event, the egg drop competition, in which students built a container to keep an egg from breaking when dropped from the top of the gymnasium.

The games were scheduled throughout the day of the expo. The egg drop event was not only the most popular with students, it was also the most popular with expo visitors.

Teacher Demonstrations

The demonstrations gave teachers a chance to show visitors what they do in their classrooms. To demonstrate the vibration of sound waves, I set up a laser so that the beam would bounce off a mirror that was attached to the speaker of a cassette recorder. I also demonstrated hologram making, displayed posters on holograms, and showed some holograms made by students. The Earth science teacher set up a demonstration on volcanic action, and the biology teacher used posters and samples to show how to identify trees by their bark and structure. The math teachers allowed visitors to use their computers and demonstrated how they use computers in the classroom. Although the demonstrations were a relatively small part of the expo, they served to make the public aware of what types of equipment and lessons are used in the classroom. In the future, we are considering adding a computer fair that incorporates the school's business department into the event.

Commercial Exhibits

The commercial exhibits allowed students to see firsthand the connection between the business

world and their classroom activities. For example, while I was giving my laser demonstration, a local hospital showed the public how doctors use a laser to perform surgery.

To arrange for the exhibits, we first wrote letters to as many organizations and businesses as we could, both in our community and in nearby Binghamton, New York, inviting them to bring science- and math-related exhibits to the expo. We also asked for any help they could give us with the expo, such as donating prize money for the student competitions. We printed the same letter in the local newspaper. It did not take long for the positive replies to start coming in as a result of both the letter campaign and the newspaper article. The businesses were much more cooperative than we ever imagined.

A chemical company showed a video on the commercial etching process they use and brought samples of the different chemicals. Another hospital sent their emergency room personnel to give blood pressure screenings. Some plastic engineers had an exhibit on how plastics are used in daily life. In addition, several environmental and conservation groups brought exhibits.

There was a great deal of interaction between the public and the exhibitors throughout the day. We encouraged the exhibitors to present students' prizes for the various events, which most did.

The exhibitors seemed to enjoy the expo as much as we did. A local company that makes team responders like those used on television game shows set up an exhibit with a scholastic bowl format. The exhibit became so popular that the company ran out of questions, so they set up teams of students with one group making up the questions and the other group answering. The company representative had a great time with the kids and was eager to return for the next expo. In fact, when I went around to personally thank each exhibitor toward the end of the expo, most were enthusiastic about coming back.

I had three goals when I started the expo project: to promote math and science, to show students the connection between the classroom and business and industry, and to get as many students as possible involved.

Teamwork Pays

By its very nature, the expo had to be a team effort. We began planning in October, and a large share of our monthly meetings revolved around the expo. The science and math teachers all worked very hard for several months, planning, helping students, and taking care of countless details until the expo took place in mid-March. In addition, individual teachers spent a lot of time writing letters, contacting organizations, collecting materials, organizing events, and helping students with their projects.

Learning Outside the Classroom

I had three goals in mind when I started the expo project: to promote science and math in our school system, to show students the connection between what we do in the classroom and what goes on in business and industry, and to get as many students as possible involved in some hands-on science beyond the classroom. Although every aspect of the expo exceeded my greatest expectations, achieving the last goal was the most gratifying part for me.

Many students engaged in quality research and learning in preparation for the event. I was especially pleased when two of my less-than-dedicated physics students actually asked for books to help with their projects. Two girls in chemistry class spent two weeks of study halls in the lab testing the acidity of foods. Two boys from chemistry worked very hard to find a way to show how bonding occurs between half-filled orbitals with oppositely spinning electrons. Even the research that students put into building the egg drop containers was absolutely fantastic.

Step Right Up to the Science Carnival

Doug Cooper

Giggles and laughter ring through the hallways, the odor of popcorn drifts to your nose, colorful booths delight your eyes, and the tingling notes of music fill the air. It is time for the annual school carnival. But this year's carnival has a different twist—it's a science carnival! It's a time and place where children and adults can share in the excitement of the carnival atmosphere and participate in doing science together.

How can you get started if you want to hold a science carnival? It's easier than you think.

Creating a Carnival

Planning is the key to a successful event. In many communities, the parent-teacher organization may already have a working model for planning and holding an annual school carnival that can be adapted for your science carnival.

Start by selecting the date to hold the carnival and then work backward from that date to create your planning timeline. Nine to 12 months of planning, committee work, and research and development may seem like a long time, but it gives everyone the opportunity to accomplish the numerous tasks.

The carnival planning process should include

- creating a budget;
- coordinating with the school faculty, administration, and maintenance departments;
- advertising; and
- setting up committees for carnival setup, cleanup, booth design, logistics and layout of the carnival space, booth construction, volunteers, booth management, security, refreshments, decorations, prizes, and tickets.

If the carnival will be a schoolwide event, then activities must be designed for all age ranges—in elementary schools, this often means kindergarten through fifth or sixth grade. Many elementary students have preschool-age brothers and sisters who might attend the science carnival, so a few activities that can also be done by very young children should be included.

Involving teachers and students in the planning and presentation of the carnival can make this a school community event. Parents also play a vital role in the carnival planning and coordination. While working on committees, parents' invaluable contributions of their ideas, time, and talents help create a fun event for the entire school.

Also, parents and adults that get involved during the science carnival have an important

role. By participating in the carnival games, they become role models for children, showing that science is for everyone. Their ability to assist younger children with the carnival games and to share their own ideas and observations sends a message to the children that even adults must try different ways of doing something.

Gathering Ideas

To gather ideas for carnival booths, start by examining traditional carnival games held in previous school carnivals or the games found along the midway at an amusement park. In some cases, parents and adults may undertake the task of discovering the science concepts involved in these games and deciding which variables may be manipulated by the carnival participants. In other cases, teachers and students can work together to examine these games.

Studying these games from a scientific perspective and identifying the scientific phenomena associated with the games may lead to putting a new twist on the activity—one that allows participants to explore science concepts, experiment with variables, solve challenges, and have fun!

Traditional Booth Redesign

An analysis of the science concepts embedded in traditional carnival booths and games is an excellent class project. Classes could analyze and redesign booths, but it is also interesting for two classes to redo the same booth, coming up with two different approaches to the science embedded in the same game. Here's how two classes could redo the same game.

A traditional game in many carnivals is a coin-toss activity: the player tosses a coin or washer and tries to get it to land on the smooth surface of a plate. Friction is one of the scientific principles of this game. If you've tried to toss a coin onto a slick dinner plate, you know that there is not much friction to slow the coin.

One class might alter the game by allowing participants to use simple materials (such as masking tape) to alter the coin to increase its friction. Another class might take a different approach to this game—the lack of friction becomes an advantage to the player. The goal may be for the player to line up two, three, or four plates in a straight line. With this approach, the tossed coin must slide along the surface of each plate and come to a rest on the last plate. To achieve this goal, participants can adjust the distance between the plates, as well as how they toss the coin.

Science Topic Booths

Another way to create booths for your science carnival is to have each class construct a booth related to one of the science topics they have been studying throughout the year. This method provides an opportunity for students to teach someone else and to demonstrate learning in a different way.

For example, one third-grade class might have studied liquids, density, and sinking and floating. So, they design a booth where participants create small boats from one piece of aluminum foil. The participants design a boat that must be 15 cm long on one side and has a small sail. When blown by a house fan, the boat must be able to carry a load of several marbles from one side of a small wading pool to the other. Another challenge might be to design a boat that will carry the most marbles. The class members, who have become the experts through the study of the subject matter, can take turns operating the booth.

More Booth Ideas

The following booth ideas can also be used at the science carnival. The instructions are directed to the booth designer(s)—who can be students, teachers, parents, and/or other adult participants.

Make Sense: What Do You Feel, Hear, Smell? This booth idea takes advantage of the human senses. Take a box with a removable lid and cut two holes in the side of the box; make the holes large enough to accommodate a person's hand and forearm. Place common objects, such as a pine cone, egg carton, or flashlight battery, inside the box. Create several different boxes filled with different objects. Participants put their arm through the holes and try to identify the objects using only their sense of touch.

To create the "hearing" part of the booth, tape record several ordinary sounds, such as water running from a faucet, a clock ticking, or popcorn popping. In the booth, play the tape and have the participants guess each sound.

To make the "smelling" activity, soak balls of paper towels with common scents such as peppermint oil, clove oil, or vinegar. Then put each type of paper towel in the bottom of its own container (mustard and ketchup squeeze bottles work well). Write the name of each scent used—and several not used—on a chart. Participants squeeze the bottle to release the scent and try to match the smell with the name of the scent.

Paper Airplane Designers. For this activity, create two or three different challenges, or targets. A large hula hoop hanging from the ceiling, a large square of colored paper taped to the floor, and a big cardboard box make good targets. Participants create paper airplanes that can be tossed at each target. The goals are to fly the plane through the hoop, land the plane on the paper square, and land the plane in the big box. Ask participants, "Can one airplane do all three tasks? Does the plane need to be redesigned for the different tasks?"

Missing Musical Jar. Create this booth by filling bottles with water and adjusting the water level of each bottle to sound a certain tone when tapped with a small mallet, which can be made by gluing a large wooden bead to a wooden dowel or pencil. (Tinkertoys also work well.) A simple melody, such as "Mary Had a Little Lamb," can then be played by tapping the bottles. Using a permanent marking pen, mark the bottles with letters to identify the tones and write out the correct tapping sequence that plays the tune on the bottles. Mark the water level of all the bottles except one, which is emptied. This empty bottle represents a "missing" musical note. Participants must experiment by varying the amount of water in the empty bottle to create the missing musical note.

Down with Gravity. This activity uses empty soda cans that have about 2.5 cm of plaster of paris in the bottom for weight. Stack three cans (with the plaster of paris at the bottom) in a pyramid—two on the bottom and one on the top. Standing about 6 m from the pyramid, participants throw a softball at the stack of cans and try to topple all three.

The participants try again, but this time, the cans are stacked with the plaster of paris weight at the top. Ask participants, "Is it easier or harder to knock the three cans over?"

Apple Basket Physics. In this activity, participants toss a softball from 6 m and try to get the ball to stay in the bottom of a wooden bushel basket turned on its side at about a 45° angle. The springy bottom of these baskets makes it a challenge to get the ball to stay in the basket. Ask participants, "What could be done to soften the landing area for the ball?" (Perhaps a soft towel placed in the bottom of the bushel basket?)

Come One, Come All

Like the traditional school carnival, making the booths colorful and festive will add to the carnival atmosphere and create an environment where parents and their children can manipulate variables as they investigate scientific phenomena together. So, step right up to the science carnival in your school!

Resources

Barber, J., and Willard, C. (1992). *Bubble Festival.* Berkeley, CA: Lawrence Hall of Science.

Beals, K., and Willard, C. (1994). *Mystery Festival.* Berkeley, CA: Lawrence Hall of Science.

Doherty, P., and Rathjen, D. (1991). *Exploratorium Science Snack-book.* San Francisco: Exploratorium Teacher Institute.

Gryczan, M. (1993). *Carnival Secrets.* Colorado Springs, CO: Picadilly Books.

Meador, J. (1987, Summer). Stalking the big dog. *The Exploratorium Quarterly,* 11(2), 18–21.

Science Festival Fun: A Teaching and Learning Experience

Verilette Parker and Brian Gerber

"Watch me eat cytoplasm," exclaims Levell as he licks the icing off a cell model made out of cookies. Levell is participating in a science festival at Camp Relitso, a summer academic enrichment program in rural Georgia. By assembling edible representatives of cell parts onto a cookie base, Levell can engage in hands-on learning of cell functions.

The cookie cell model is just one of many interactive exhibits at Camp Relitso. Unlike a science fair that consists of static displays of science investigations completed by students, a science festival includes interactive science exhibits that students share with one another.[1] As students manipulate interactive exhibit materials, their peers answer questions and discuss pertinent science concepts. At Camp Relitso, students take part in a five-week program that prepares them for the culminating festival. During this time, seventh and eighth graders study a life science unit on the human body systems, while fifth and sixth graders study a physical science unit on matter. Both classes employ the learning cycle as the primary teaching approach.[2,3] In this format, students explore simple manipulative materials, discuss science concepts, and apply newly acquired knowledge and skills to different science tasks.

The following article is geared toward helping you develop your own science festival. Why not encourage your students to develop their own science festival these last few weeks before summer break? They may just have fun.

Science Festival Setup

Students participating at the science festival at Camp Relitso engage in science activities according to the learning cycle. Inquiry-based learning experiences are consistent with the *National Science Education Standards* and provide cooperative learning opportunities for students.

Our science festival lasts approximately one hour: The seventh- and eighth-grade students spend the first 30 minutes sharing their interactive life science exhibits with the fifth- and sixth-grade students. After the older students guide the younger participants through the life science exhibits, the younger students guide the older students through the physical science exhibits. Most students are able to complete all of the exhibit activities within the time allotted and even

have time for return visits to their favorite ones. Many of the interactive exhibits at Camp Relitso are based on activities from *Using the Learning Cycle to Teach Physical Science* and *Science Discovery Works*.[4,5] Popular exhibits include Balloon Blow-Up, Vibrations and Pulsations, What a Feeling, Floating Eggs, Colorful Liquid Layers in a Straw, Acid Testing, Silly Matter, Pretend Fossils, and Electric Circuits.

During the festival, students set up their exhibits on individual tables so that festival participants can easily observe and manipulate the materials. As participants "play" at each exhibit, student exhibitors ask questions and share information. Before long, students guide one another through active learning experiences. What's more, students experience dual responsibilities—teaching and learning. First, student exhibitors become teachers who guide participants through interactive exhibits to impart science knowledge and skills. Then, students switch roles, and the exhibitors become the participants.

Popular Interactive Exhibits

Each year, students create the interactive exhibit Balloon Blow-Up, which is an easy activity that students really enjoy. All that's required are two-liter soda bottles, vinegar, baking soda, balloons, funnels, and measuring cups and spoons. Depending on class size, you will need one to two dozen bottles; a local recycling company donated a box of recycled bottles, which were cleaned prior to the festival. With the help of the exhibitors, participants follow written instructions as they carefully measure and combine the proper amounts of baking soda and vinegar. The faces of the participants are pensive as they measure, pour, and mix the ingredients. Their eyes widen as the balloons inflate.

When baking soda and vinegar are mixed, the soda bottle fills with a gas that inflates the balloons. Students learn that a chemical reaction has taken place, producing a new substance. In this case, the new substance is the gas carbon dioxide.

The Gumdrop Molecule Model exhibit is a great way to help students visualize how atoms combine to form molecules. For this exhibit, students refer to a picture of the carbon dioxide molecule to make tangible models of carbon dioxide molecules. Students make their models by inserting two toothpicks into a red gumdrop, which represents the carbon atom, and adding two green gumdrops to the ends of the toothpicks, which represent the oxygen atoms. As students make their models, exhibitors explain the chemical makeup of carbon dioxide and other elements on the periodic table.

For the interactive exhibit Floating Eggs, students examine the relationship between salinity and density. Students will need two 16-ounce, clear jars; water; salt; and a measuring spoon. Have exhibitors fill both jars with water but add salt (approximately five teaspoons) to only one jar. They should stir the solution until the salt has completely dissolved. As the participants approach, exhibitors challenge them to explain what they are about to witness: The egg placed in the first jar sinks, while the egg placed in the saltwater floats.

In another density exhibit, Colorful Liquid Layers in a Straw, students are challenged to layer saltwater solutions. To set up this exhibit, have students prepare four solutions, each with a different salinity, and then add food coloring to each solution. (I recommend the following ratios of salt to water, ranging from heaviest to lightest: 1 quart water to 1 cup salt; 1 quart water to 2/3 cup salt; 1 quart water to 1/3 cup salt; 1 quart water to no salt.)

As students try to make colorful liquid layers in clear, plastic straws, they discover the arrangement of the colored solutions has to be in a particular order. (The solution having the least density must be at the top, while the solution having the greatest density is at the bottom.) The colors will blend together if students layer the solutions incorrectly.

Blue Litmus Paper for Acid Testing is always an enticing exhibit. Students use strips of blue litmus paper to determine the acid levels of everyday household items. Armed with pipettes, students place drops of vinegar, coffee, milk, and orange juice onto different strips of litmus paper. Right before their eyes, the litmus paper turns pink. Students also test the pH of ammonia, a sugar solution, and a baking soda solution for comparison purposes. It's wonderful to hear students discussing chemistry terms such as acids, bases, and indicator.

In a similar exhibit, students use Bromothymol Blue (BTB) for acid testing. In this activity, students will need BTB, vinegar, ammonia, and pipettes. Students must wear goggles, aprons, and gloves while doing this activity. Using the pipettes, students place drops of vinegar into vials containing BTB until they observe a color change from blue to yellow. Then, they add ammonia to the yellow BTB, drop by drop, until it becomes blue again. This time, students learn that BTB is an indicator for both acids and bases.

Learning about states of matter is always fun, especially when creating Silly Matter. Students will need cornstarch and water (ratio of three parts cornstarch to two parts water works well). Have students mix the ingredients together to make a pliable substance that can be rolled into a ball. At first the substance is pliable like Play Dough but soon changes to a liquid. Shortly after students roll the substance into a ball, it starts to ooze through their fingers. Exhibitors then describe the properties of solids and liquids.

At the Pretend Fossils exhibit, students pretend that chocolate chips in cookies are fossils of prehistoric animal bones embedded in clumps of dirt. Students learn that substances within mixtures can often be removed by mechanical means. Students simulate this by using toothpicks to remove the chips from the cookies as archaeologists remove bones from strata.

The exciting Electric Circuits exhibit allows students to explore the parts of electric circuits for a hands-on investigation of conductors and electricity. First, exhibitors display the necessary materials to form circuits. With the help of the exhibitors, students successfully close the circuits, lighting bulbs and sounding buzzers. Then, copper and aluminum wires are inserted between the battery and light bulb to show that both metals are conductors.

In Vibrations and Pulsations, participants are given a small ball of clay (about the size of a marble), which they flatten on the palm side of their wrist. They then insert a match into the clay in an upright position. Students observe the vibration of the match that occurs from the pulsation of the blood in the body caused by the heart pumping blood through blood vessels. They count the number of vibrations per minute, which for children is generally between 80 and 140.

What a Feeling science can give students! At this activity station, participants test the sensitivity of their skin. Exhibitors bend a large paper clip to form a symmetrical U shape, so that both ends of the paper clip are even with each other. Festival participants are asked to look away as the exhibitor touches their upper arm, lower arm, thumb, and finger with the paper clip. The par-

ticipating students are to determine whether one or two points of the paper clip were pressed on each part of their body. Because there are not as many nerve endings in the upper and lower arm as in the thumb and finger, the participants' responses will be more accurate when the paper clip touches the thumb and finger.

Camp Relitso science festival participants tell me that it's fun doing science. As one student remarked, "At first I was scared to share my exhibit, but then I got excited when others started playing with the materials and asking me questions." In addition to being a successful teaching and learning experience for students, a science festival can be an enjoyable way to review the entire year in an hour!

References

1. Parker, V. 1996. Family fun at SCI-FEST. *Science and Children* (34)2: 24–25, 50.
2. Gerber, B.L. 1996. *The relationship among informal learning environments, teaching procedures, and scientific ability.* Unpublished doctoral dissertation, University of Oklahoma, Norman.
3. Renner, J.W., and E.A. Marek. 1990. An educational theory base for science teaching. *Journal of Research in Science Teaching* (27)3:241–46.
4. Beisenherz, P.C., and M. Dantonio. 1996. *Using the learning cycle to teach physical science.* Portsmouth, N.H.: Heinemann.
5. Badders, W., L. Bethel, V. Fu, et al. 1996. *Science discovery works.* Parsippany, N.J.: Silver Burdett Ginn.

Appendices

Resource List

BOOKS

Bochinski, J. B. 1996. *The Complete Handbook of Science Fair Projects, Rev. Ed.* New York: John Wiley and Sons. ISBN: 0-471-12378-1. Grades 7–12.

Written by a former science fair winner and judge, this book includes 50 award-winning projects from actual science fairs, plus a list of 400 workable science fair project topics. Also provides the latest information on alternative science fair competitions.

Bombaugh, R. 1999. *Science Fair Success, Rev. and Exp.* Berkeley Heights, NJ: Enslow Publishers. ISBN: 0-7660-1163-1. Grades 6–12.

This guide takes students through all the steps required for a science fair project, from selecting the topic to final presentation. High-quality illustrations, Web references, writing tools, suppliers, and 125 prize-winning projects.

Fredericks, A. (Introduction by Isaac Azimov.) 1991. *The Complete Science Fair Handbook: For Teachers and Parents of Students in Grades 4–8.* Santa Monica, CA: Goodyear Publishing. ISBN: 0-673-38800-X. Grades 4–8.

"Dr. Asimov and I," writes the author, "were concerned that too many science fairs featured a host of solar system models and an over-abundance of volcanoes...." This book suggests many alternative topics (by grade level, though no actual project examples) and includes checklists, time lines, and how-tos of running a science fair.

Henderson, J., and H. Tomasello. 2001. *Strategies for Winning Science Fair Projects.* New York: John Wiley and Sons. ISBN: 0-471-41957-5. Grades 6–12.

This student-oriented book covers all aspects of participating in a science fair, including selecting a research topic, using the scientific method, anticipating problems, safety issues, and dealing with difficult judges.

Iritz, M. H. 1999. *More Blue Ribbon Science Fair Projects.* New York: McGraw-Hill. ISBN: 0-071-34668-6. Grades 4–8.

This updated guide on science fair projects, designed for first-timers, comes complete with a built-in science project organizer—checklist, planner, and log—that guides teachers and students step-by-step through the entire process. Gives advice on how to turn any idea into a full-fledged, affordable project and presents the inside scoop on exactly what judges are looking for.

Perry, P. J., and D. Ellinger (illus.). 1995. *Getting Started in Science Fairs: From Planning to Judging.* New York: Tab Books. ISBN: 0-070-49526-2. Grades: elementary.

This book offers advice on science fair basics and presents tested science projects. Numerous aspects of science fairs are covered, including science fair objectives, motivational techniques, and the role adults should play.

Phillips, G., L. Hoffman, and J. Armbrust (illus.). 1997. *Middle School Science Fair Projects.* Redding, CA: Instructional Fair. ISBN: 1-56822-429-X. Grades 5–8.

Projects cover a wide variety of experimental topics. Students conduct their own research and record their data on charts and graphs provided. Information for ordering hard-to-find materials is provided.

Rosner, M. A. 1999. *Science Fair Success Using the Internet.* Berkeley Heights, NJ: Enslow Publishers. ISBN: 0-7660-1172-0. Grades 6–12.

This book provides guidelines for on-line communications, such as forums, message boards, newsgroups, and e-mail. Students who follow the guidelines will learn how to reach scientists online (websites are listed) and get answers to complex questions from experts. Also, mailing lists, "e-zines," and file-transfer protocol (FTP) are mentioned as sources of information for research.

VanCleave, J. 2000. *Janice VanCleave's Guide to the Best Science Fair Projects.* New York: John Wiley and Sons. ISBN: 0-471-32627-5. Grades 3–6.

The author demonstrates how to plan ahead and to turn simple experiments into competitive science fair projects. A teacher resource, since it is written at a post-elementary reading level.

Vecchione, G. 1997. *100 Amazing Make-It-Yourself Science Fair Projects.* New York: Sterling Publications. ISBN: 0-806-90367-8. Grades 6–12.

This collection of experiments, from building a working telescope to creating a miniature ecosystem for display, can be carried out with a minimum of special materials. Each project demonstrates a different important principle.

Wee, P. H. 1999. *Science Fair Projects for Elementary Schools.* Lanham, MD: Scarecrow Press. ISBN: 0-8108-3543-6. Grades 2–5.

This book gives step-by-step instructions for hands-on science fair projects. One section provides parents, teachers, and librarians with sample letters, forms, and layouts to facilitate setting up the science fair.

WEBSITES

Steps to Prepare a Science Fair Project
www.isd77.k12.mn.us/resources/cf/steps.html

This site, called "Cyber-Fair: A Resource for and by Elementary Students," includes an ideas list and sample student projects.

Science Fair Homepage
www.stemnet.nf.ca//~jbarron/scifair.html

Provides links to many other sites, including "Science Fair Projects: A Resource for Students and Teachers."

Science Fair Workshop
www.eduzone.com/tips/science/showtip4.html

Emphasis on links that explain how to create a display.

Discovery Channel School: Science Fair Central
http://school.discovery.com/sciencefaircentral/

"A comprehensive guide to creating your science fair!" Includes handbook, ideas, links, tip sheets.

The Kids Guide to Science Projects
http://edweb.tusd.k12.az.us/jtindell/

"A science experiment is nothing more than a way to solve a problem. These pages have been created to give you some ideas and resources, show you how to start, and take you step-by-step through the scientific process."

The Ultimate Science Fair Resource
http://www.scifair.org/

A variety of resources and advice.

The Exploratorium Learning Studio—Science Fairs
studio.exploratorium.edu

Gives students, teachers, and parents advice for locating resources and supplies. The Learning Studio is part of the Exploratorium (the world-renowned "hands-on" museum of science, art, and human perception in San Francisco). The Learning Studio provides electronic, multimedia, and print-based learning resources.

Kids Involved Doing Science/Family Science Night Handbook
www.kids.union.ed then go to "Family Science Night Handbook"

An online handbook designed to help a team of four to six adults conduct an elementary school Family Science Night. Includes information on how to plan and present such an event, 25 activities, and sources of inexpensive supplies.

NSTA Contests

Besides competing in traditional science fairs, students can enter projects in national contests. The National Science Teachers Association (NSTA), with major support from Sears, Roebuck and Co. and Toshiba, administers two student competitions—the Craftsman/NSTA Young Inventors Program and the Toshiba/NSTA ExploraVision Awards Program. Both programs award U.S. Savings Bonds to winners and a free gift to every student who enters.

Craftsman/NSTA Young Inventors Program

This annual competition, funded by Sears, Roebuck and Co. and administered by NSTA, challenges students to use their creativity and imaginations along with science, technology, and mechanical abilities to invent or modify a tool. The tool must perform a practical function—for example, the tool might be one that mends, makes life easier or safer in some way, entertains, or solves an everyday problem.

Each student-entrant must have one teacher/adult advisor who is an elementary classroom, science, technology, or special education teacher. The adult advisor may also be a boys' or girls' club leader or member of a similar organization.

The two national winners—one from grades 2–5 and one from grades 6–8—each receives a $10,000 U.S. Savings Bond. Ten national finalists (five from each grade category) each receives a $5000 U.S. Savings Bond. The teacher advisors and schools of the twelve finalists each receives prizes from Sears, Roebuck and Co.

Next entry deadline: March 4, 2003

For entry materials and/or a teacher's guide, call 1-888-494-4994 or e-mail the program at *younginventors@nsta.org*. Visit the website: *www.nsta.org/programs/craftsman*

Toshiba/NSTA ExploraVision Awards Program

This annual competition, funded by Toshiba and administered by NSTA, invites students of all skill and ability levels to consider the impact that science and technology have on society and how innovative thinking can change the future. Students compete in teams at one of four grade-level categories—K–3, 4–6, 7–9, and 10–12. The teams explore an existing technology, then research and describe a scientifically sound use for the technology as it might exist in twenty years. A teacher-mentor and optional community advisor facilitate the team's effort.

Each student on the four first-place teams wins a $10,000 U.S. Savings Bond; each student on the eight second-place teams receives a $5000 U.S. Savings Bond. Canadian winners receive Canada bonds purchased at comparable issue price.

Next entry deadline: February 4, 2003

For entry materials and to view past winners' projects, visit the program's website: *www.toshiba.com/tai/exploravision*. Send an e-mail request to *exploravision@nsta.org* or phone toll-free (800) EXPLOR9.

Appendix C

List of Contributors

(in alphabetical order)

Lawrence J. Bellipanni, lead author of both "What Have Researchers Been Saying about Science Fairs" and "In the Balance," continues as an associate professor of science education at the University of Southern Mississippi.

Stephen C. Blume, currently assistant director of federal programs for the St. Tammany Parish Public Schools, Covington, Louisiana, was an elementary science curriculum specialist when he wrote "Scientific Investigations."

Ruth Bombaugh was a seventh- and eighth-grade science teacher at Langston Middle School, Oberlin, Ohio, when she wrote "Mastering the Science Fair." She is now an assistant professor of science education, middle school, and urban education at Cleveland State University.

Cheryl Cook, one of the authors of "Science Fairs for All," continues as a K-1-2 multi-age classroom teacher at North Star Elementary School in Anchorage.

Doug Cooper was a teacher education supervisor and teacher inservice provider at Pacific Science Center in Seattle, Washington, at the time he wrote "Step Right up to the Science Carnival."

Cecelia Cope, author of "Science Fair Fatigue," continues as a middle school science teacher at Christ the King/St. Thomas the Apostle School in Minneapolis.

Donald R. Cotten, now dean of the Graduate School at the University of Southern Mississippi, was an associate professor of science education at that university when he contributed to "In the Balance."

Charlene M. Czerniak, author of "The One-Hour Science Fair," is a professor in the Department of Curriculum and Instruction—Elementary and Middle Grades Science Education at the University of Toledo, Ohio.

Deborah Fort is a Washington, D.C. freelance writer and editor specializing in science, education, and women's issues. She was a parent volunteer at the science fair she describes in her article, "Getting a Jump on the Science Fair."

Gail C. Foster was a science teacher at the Energy Management Center, Port Richey, Florida, at the time she wrote "Oh No! A Science Project!" Currently, she is an independent educational consultant in Florida. She develops environmental and science instructional materials for students and teachers and conducts professional staff development.

Brian Gerber, now an associate professor in the Department of Secondary and Middle Grades Education at Valdosta State University, was an assistant professor at that university when he contributed to "Science Festival Fun: A Teaching and Learning Experience."

Jan Marion Kirkwood, a contributor to "In the Balance," continues as a science teacher at Lewis Middle School, Natchez, Mississippi.

James Edward Lilly, co-author of "What Have Researchers Been Saying about Science Fairs?", is a professor of science at Xavier University of New Orleans, having been a doctoral candidate at the Center for Science and Mathematics Education at the University of Southern Mississippi at the time the article was written.

Margaret McNay is an associate professor on the Faculty of Education, University of Western Ontario, London, Ontario. At the time she wrote "The Need to Explore: Nonexperimental Science Fair Projects," she was an assistant professor in the Department of Early Education, The University of Alberta, Edmonton.

Verilette Parker, now an assistant professor in the Department of Early Childhood and Reading at Valdosta State University, was a teacher facilitator at Lowndes Middle School in Valdosta, Georgia, when she was the lead author of "Science Festival Fun: A Teaching and Learning Experience."

Teralyn Ribelin, one of the authors of "Science Fairs for All," continues as a K-1-2 multi-age classroom teacher at North Star Elementary School in Anchorage.

Susan Shaffer (Iannacci), who wrote "Prepare for Science Fair," continues to teach middle school science and serve as the science department chairperson at Ridley Middle School in Ridley Park, Pennsylvania.

Donna Gail Shaw, lead author of "Science Fairs for All," continues as a professor of elementary science education at the University of Alaska-Anchorage.

Debbie Silver, who wrote "Science Fairs: Tired of the Same Old, Same Old?", is a retired teacher and professor. She is currently a speaker and consultant and the author of *Drumming to the Beat of a Different Marcher* (Incentive Publications, 2002), living in Clinton, Mississippi.

Linda H. Sittig, author of "Whoever Invented the Science Fair...," continues as a reading specialist for the Fairfax County, Virginia, Public Schools and, in addition, is now an associate professor at Shenandoah University.

Norman F. Smith, author of "Why Science Fairs Don't Exhibit the Goals of Science Teaching," is a mechanical and aeronautical engineer and a former NASA aerospace research scientist. The author of several trade books for young people, he lives in Shelburne, Vermont.

John Stiles has been an educator for over thirty years. A doctoral candidate in science education at the University of Iowa in Iowa City at the time he wrote "The Trouble with Science Fairs," he has taught at the elementary through university graduate levels. He is currently high school principal at Rose Marie Academy in Nonthaburi, Thailand.

Evelyn Streng was an associate professor at Texas Lutheran University when she wrote "Science Fairs? Why? Who?" Now Professor Emerita, she continues to teach in many capacities, including at a Texas Elderhostel, and acts as director of the Fiedler Memorial Museum at Texas Lutheran University.

Daniel Wolfe, author of "Go for an Expo!", retired in 2000 after 35 years of teaching chemistry and physics at Susquehanna Community High School, Susquehanna, Pennsylvania. He is currently doing grant writing for a local hospital and is a volunteer in a food recovery and distribution center in Binghamton, New York.